TITRIMETRIC METHODS

TITRIMETRIC METHODS

Proceedings of the Symposium on Titrimetric
Methods held at Cornwall, Ontario, May 8-9, 1961

Sponsored by
The Chemical Institute of Canada
Analytical Subject Division, Cornwall District Section

Edited by
Dr. D. S. Jackson

A publication of
The Chemical Institute of Canada

SPRINGER SCIENCE+BUSINESS MEDIA, LLC
1961

ISBN 978-1-4899-2704-0   ISBN 978-1-4899-2702-6 (eBook)
DOI 10.1007/978-1-4899-2702-6

Library of Congress Catalog Card Number 61-17728

CONTENTS

ELECTROMETRIC TITRATIONS

COMPLEXOMETRIC TITRATIONS

TITRATIONS IN NONAQUEOUS SOLVENTS

Electrometric Titrations

AN AUTOMATIC "WET" CHEMICAL ANALYZER FOR PROCESS MONITORING AND CONTROL

Walter J. Harrison
*Fisher Scientific Company,
Pittsburgh, Pennsylvania*

ABSTRACT

An instrument has been developed which can sample a process stream, prepare the sample for analysis by the addition of required reagents, titrate the mixture to an electrically determined end point, record the results in either digital or analog form, dispose of the sample, and repeat the cycle indefinitely.

An output signal can be made available for use in process control or to operate high or low concentration alarms.

Time required for a complete analysis cycle varies between 3 and 10 min depending on the complexity of the analysis.

Normal precision expected with the instrument is a standard deviation of 0.01 ml of titrant. Much better results are possible under certain circumstances.

Successful results have been obtained with both simple and complex analytical problems, i.e., mercaptan sulfur in gasoline and "Bayer liquors."

The instrument is designed so that by the selection of the proper plug-in units and the proper programming almost any volumetric laboratory procedure can be duplicated with better precision and accuracy than is normally obtainable in the laboratory.

3

INTRODUCTION

In recent years much work has been done in the chemical industry to completely automate various chemical processes. Computers, both analog and digital, have been applied to this problem with varying degrees of success. In order to control a process, however, information concerning the state of the process at any given time must be available, and available in such a form that it can be readily utilized. The vast majority of instrumentation in use in the chemical plant today depends solely on physical measurement, i.e., temperature, pressure, viscosity, rate of flow, etc. Instruments which are capable of measuring concentrations of substances used in the process, analytical instruments, still depend on physical measurements for results, i.e., infrared and ultraviolet analyzers, pH meters, gas chromatographs, etc. Many problems of process monitoring, however, do not lend themselves to these techniques and as a result chemical analysis must be performed in a control laboratory. This makes complete automation of a process impossible. Until continuous automatic chemical analysis is possible, the only approach available is to automate the control laboratory function and feed the information directly to a device, whether it be a computer or simple pneumatic controller, which in turn can control the process in question.

An attempt to automate chemical analysis in the plant has resulted in the instrument or family of instruments described in this paper. The qualifications of an instrument of this type are many and varied, depending on the system with which it must be used, but, in general, the instrument must be:

1. Reproducible
2. Reliable and trouble free
3. As rapid as possible in the performance of a given analysis
4. Capable of producing results in a usable form

Since the techniques of analytical chemistry are many and varied and since it would be impossible to automate all of them with a single instrument, it was decided to pick a general method which would have a wide application to as many problems as possible. Volumetric analysis was chosen for the first approach and pH and potentiometric methods were attacked first. Colorimetric techniques were chosen for the second approach. Most of the work done thus far has been on pH and electrometric systems.

DESCRIPTION OF APPARATUS

Figure 1 shows the basic instrument programmed to do the simplest of all analyses, an acid—base titration. The instrument is composed of two principle sections, the electrical section and the chemical section. The electrical section contains a pH electrometer, a control panel, a bank of adjustable timers, a digital counter and printer combination, panels which contain relays, a servo-amplifier, a programming panel, a master stepping relay, and necessary power supplies. The chemical section contains a calibrated sample loop, a reaction cell, a servo-driven syringe, a stirring mechanism, and all the valves and plumbing required to handle the sample, titrant, and diluent.

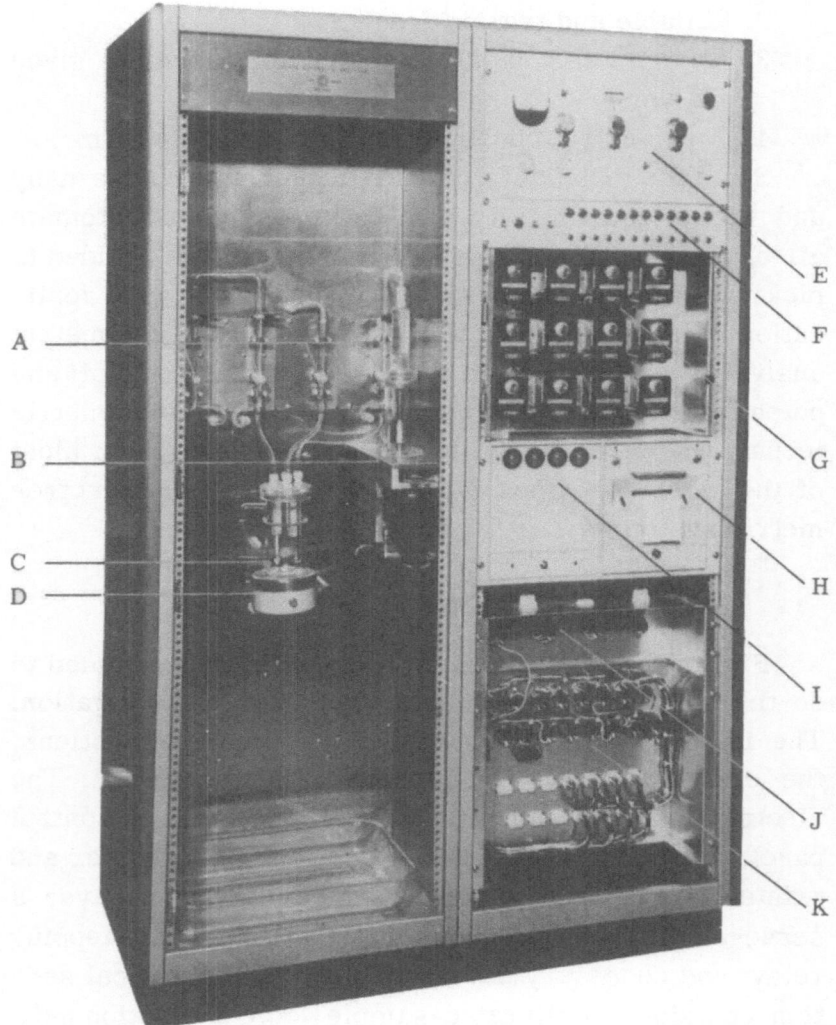

Fig. 1. Fisher Automatic Analyzer. A) Motor valves; B) syringe assembly;
C) reaction vessel; D) stirring assembly; E) electrometer; F) control panel;
G) timer panel; H) printer; I) counter; J) electronic relay panel; K) relay
panel.

THEORY OF OPERATION

To perform any volumetric analysis several basic things are required.

1. A sample must be removed from the process stream and measured to a precise volume.
2. Any required reagents, such as complexing agents, must be measured to a precise volume and mixed with the sample.
3. A titrant of known concentration must be added to this mixture until an equivalence point is reached.
4. The equivalence point must be detected and the volume of titrant or a function of the titrant volume must be recorded.
5. The sample mixture must be discarded and the process repeated.

With these basic requirements in mind, it was decided to build an instrument which would operate in discrete steps.

A stepping relay with twelve operational positions was chosen for the basic control unit. This stepping relay was connected so that it could be controlled either by push buttons for manual operation or by a combination of timers and detection devices which would signify an event successfully completed in a given step. Seven control points were made directly available in any step but any number of control points are available by the addition of multiplexing relays.

Two methods of selecting and introducing the sample were devised. For small samples, 0-10 ml, a calibrated sample loop, located between the motor-operated three-way valves, was used. A branch of the process stream was continually circulated through this loop and at the time of sampling both valves were actuated to trap the sample. By further rotation of these valves the sample was flushed into a reaction cell by a suitable solvent.

For larger samples, a modified pipet design was utilized. The branch of the process stream was allowed to flow through a self-zeroing type pipet which was controlled by a motorized three-way valve, and at the time of sampling the valve was actuated to cut off the stream and allow the sample to drain by gravity into a reaction cell.

The measuring of reagents was accomplished in the same manner as described for a large sample, with the exception that as the pipet overflowed either a pair of contacts or a light-photocell combination located in the overflow line actuated a sensitive relay, which in turn caused the valve to shut off the flow. The photocell arrangement was used only in the case of nonconducting liquids.

The actual titration of the sample was performed by means of a servo-driven syringe. This unit requires some detailed explanation. Figure 2 shows the syringe unit together with one of the data systems employed. The syringe was designed along the principles described by Lingane [1] but the construction was quite different. A 40-thread/in. precision lead screw drives a nylon nut which is prevented from turning by a pin which rides in a milled slot in the plastic case. A single ball bearing which has a hole through its axis is held in this nut by set screws. The plunger of the syringe is cemented to this bearing by means of a brass sleeve and Eastman 910 cement. The plunger is hollow and when the nut is at the bottom of its travel the lead screw extends through the hole in the bearing and up into the plunger of the syringe.

The purpose of the bearing is to make the plunger self-aligning with the barrel, which is held to the top of the unit by means of a screw cap.

The screw is driven by a two-phase servomotor through a two-speed electrically actuated transmission. This transmission allows the screw to be driven at either 100 or 25 rpm. Direction is reversed by changing phase on one winding of the servomotor.

A binary shaft encoder, which produces 100 pulses per revolution of its input shaft, is geared to the lead screw by means of a timing belt. This shaft encoder supplies its output to a beam switching counter which stores the pulses and presents the total accumulated number of pulses to the digital printer on command. With a 50-ml syringe one pulse or count is approximately

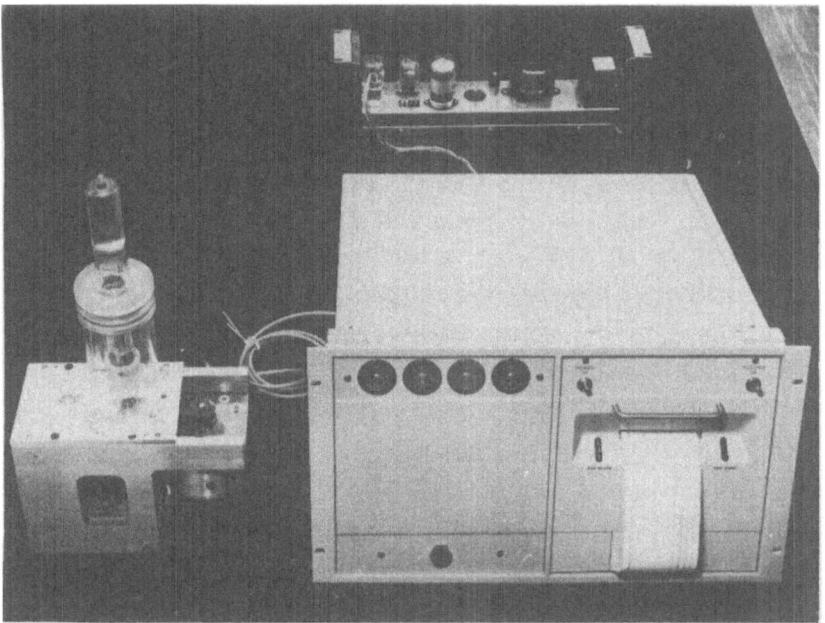

Fig. 2.

equal to 0.004 ml of titrant. The data system has been described in detail elsewhere [2].

In some cases a potentiometer has been used in place of the encoder to feed an analog signal to a strip-chart recorder for read-out. The control of the syringe unit is accomplished by a combination of devices. A three-way motorized valve is used to switch the syringe from a titrant supply to the reaction cell. Until this valve is properly positioned the unit will not function. A manual switch is provided to fill and empty the syringe, and limit switches are provided to prevent overtravel.

To remove the effects of backlash, a novel switching circuit has been employed which causes the nut to be driven up slightly from the lower-limit switch after the filling operation, thus taking up any slack in the mechanism. This means that each titration begins from exactly the same position of the nut with all backlash removed.

The servomotor itself receives its signal from a conventional servo-amplifier, which in turn receives a chopped error signal from a null type pH electrometer. The electrodes which supply the signal to the electrometer are located in the reaction cell. Since the electrometer is of the null-balance type, a preset end point is required and this presupposes a knowledge of the titration curve. This electrometer system has been described previously [3].

The reaction cell employed was made of glass and was provided with a Lucite cover, containing electrode and liquid inlets, which clamped on the cell. Overflow and drain ports were connected to a waste manifold by means of polyethylene or Tygon tubing.

Stirring was accomplished by a magnetic stirrer, in most cases, although overhead stirring and even ultrasonics were used in some special cases.

Solenoid valves of stainless steel, nylon, etc., were used where conditions permitted.

Since this instrument consists of many varied systems, a convenient method of tying them all to the basic control system was required. This was accomplished with a computer-type patch board. All of the control points from the stepping relay were brought to this board and all of the power input points from the various systems and units were likewise terminated there. To program a given operation in a given step, all that was required was to insert a jumper wire between the proper points on this board. Since all units such as valves, glassware, etc., were standardized, all that is required to change from one analysis to another is to select the proper number and size of units, install them, and change the program accordingly.

Figure 3 shows a flow sheet for a typical simple analysis which involves the use of two additional reagents besides the sample, titrant, and solvent. This analysis could be programmed in the following way.

1. (a) Sample is trapped in the loop and flushed into the reaction cell with water.
   (b) One reagent is added from a reagent pipet.
2. The mixture is stirred for 2 min.
3. The second reagent is added.
4. The titration is performed.
5. (a) The result is recorded.
   (b) The sample is discarded.
   (c) The syringe is refilled.
6. (a) The reaction cell is flushed with water.
   (b) Both reagent pipets are filled.
7. The wash water is drained from the cell.

Steps 8-12 are skipped and the cycle begins again. If, for example, flushing of the reaction cell were

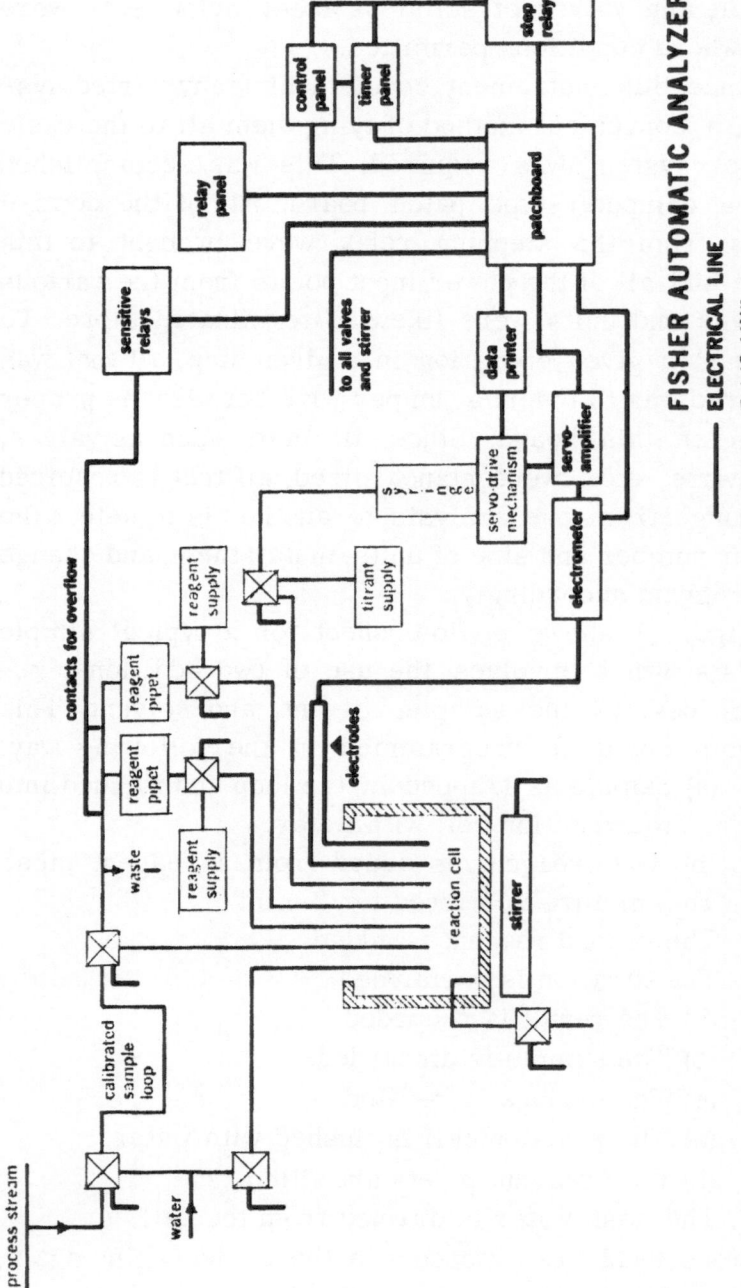

FISHER AUTOMATIC ANALYZER

——— ELECTRICAL LINE

——— FLUID LINE

Fig. 3. Schematic representation of one "task-tailored" Fisher Automatic Analyzer assembly. The same units may be combined differently for other specific applications.

not required, the following program could be used:

1. (a) Sample is trapped and flushed into the reaction cell.
   (b) The first reagent is added.
2. (a) The mixture is stirred for 2 min.
   (b) The first reagent pipet is refilled.
3. The second reagent is added.
4. (a) The titration is performed.
   (b) The second reagent pipet is refilled.
5. (a) The result is recorded.
   (b) The sample is discarded.
   (c) The syringe is refilled.

Steps 6-10 could then be programmed the same as steps 1-5, and steps 11 and 12 could be skipped.

In some complex analyses, such as the analysis of "Bayer liquor," all 12 steps were required for a single sample. This analysis involved three titrations and the addition of four reagents to each sample.

The read-out system selected for use on a given problem depends on several factors. The most accurate system is the digital system mentioned above. Digital information can of course be used to feed a control computer and also to trigger rather simple high and low concentration alarms. If recording the time at which the analysis is performed is an important factor, some type of digital clock, such as the one shown in Fig. 4, must be used. A less costly system utilizing a potentiometer and a strip chart recorder can show time directly, and a controller as well as alarm contacts can be supplied as a part of the recorder. Resolution suffers with this type of system, however, but for some applications this is perfectly adequate.

Some processes possess large amounts of inertia

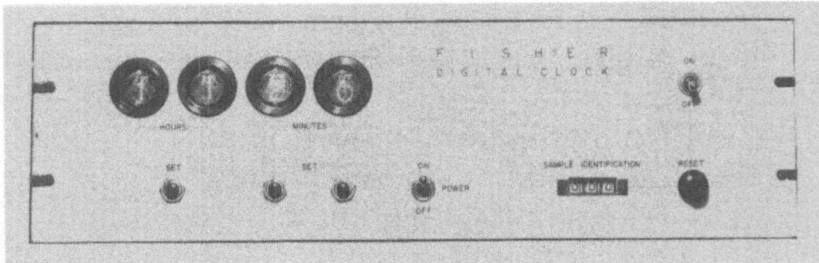

Fig. 4.

and rapid repetitive analyses serve no real purpose. In such cases it is perfectly feasible to sample the process at different points or even sample other streams with one instrument. This can be done by a sequential switching of streams to the instrument. In one case investigated it was required to sample six streams sequentially. This was done by means of an additional stepping relay which switched input valves and also supplied an identifying code number to the printed analytical result.

TYPICAL RESULTS

Many of the results obtained with this instrument have been reported elsewhere [3-6] but Tables I, II, and III show some typical results obtained on systems of varying complexity.

Table I shows results obtained by titrating 56 samples of approximately 0.05 N acetic acid with 0.01 N sodium hydroxide to a pH of 8.2. The count values were not reduced to volumes since the sample loop had not been calibrated during this run. The standard deviation obtained on these 56 samples was 1 count, or approximately 0.004 ml. A conservative 95% confidence level in this case would be ± 0.01 ml of titrant.

Table I. Titration of 0.05 N Acetic Acid with 0.01 N
           Sodium Hydroxide

Sample	Count	d^2	Sample	Count	d^2
1	762	0.09	29	763	1.69
2	760	2.89	30	764	5.29
3	761	0.49	31	759	7.29
4	762	0.09	32	761	0.49
5	762	0.09	33	761	0.49
6	761	0.49	34	761	0.49
7	761	0.49	35	762	0.09
8	762	0.09	36	761	0.49
9	762	0.09	37	761	0.49
10	761	0.49	38	764	5.29
11	762	0.09	39	762	0.09
12	760	2.89	40	762	0.09
13	762	0.09	41	763	1.69
14	761	0.49	42	762	0.09
15	762	0.09	43	762	0.09
16	762	0.09	44	761	0.49
17	762	0.09	45	763	1.69
18	762	0.09	46	761	0.49
19	763	1.69	47	760	2.89
20	762	0.09	48	762	0.09
21	762	0.09	49	761	0.49
22	762	0.09	50	760	2.89
23	762	0.09	51	761	0.49
24	762	0.09	52	761	0.49
25	764	5.29	53	761	0.49
26	763	1.69	54	762	0.09
27	763	1.69	55	762	0.09
28	762	0.09	56	761	0.49

Mean count = 761.7, σ = 1.0, Coefficient of variation = 0.13%

Table II. Titration of Gasoline and No. 2 Fuel Oil
        for Mercaptan Sulfur

Sample 1 (gasoline), ppm S as RSH			Sample 2 (No. 2 fuel oil), ppm S as RSH	
6.1	6.1	6.0	15.9	15.9
6.1	6.1	6.0	16.0	15.9
6.0	6.0	6.0	15.9	15.9
6.0	6.1	6.0	15.9	15.9
6.0	6.0	6.0	16.0	16.0
Average		6.0	Average	15.9
Standard deviation		0.05	Standard deviation	0.05

Table III. Precision Test of the Analyzer From
        Repetitive Analyses of a Spent Liquor*

Caustic, 144.7 g/liter	Carbonate, 40.6 g/liter	Alumina, 46.8 g/liter
145	42	47
144	40	47
145	40	47
144	40	47
145	40	47
144	40	47
145	42	47
145	40	47
144	40	47
144	40	47
144	40	47

Caustic, 144.7 g/liter	Carbonate, 40.6 g/liter	Alumina, 46.8 g/liter
145	40	47
144	40	47
145	42	47
144	40	47
145	42	47
145	40	47
145	40	47
145	40	47
145	40	47
145	42	47
144	40	47
145	42	47
145	40	47
144	42	47
145	40	47
145	40	47
145	42	47
145	40	47
144.5	41.1	47

Standard deviations

0.35 g/liter	0.93	0.00 g/liter†
0.03 ml	0.08 ml	0.00 ml

*An atypical liquor because of its volume having been diluted.
†Individual values are rounded to nearest unit. Therefore, a value of about 0.1 g/liter might be more realistic.

Table II shows some results obtained in titrating gasoline and No. 2 fuel oil samples for mercaptan sulfur using 0.001 N silver nitrate. These results were reduced directly to ppm of sulfur and standard deviations of 0.05 ppm were obtained with both the gasoline and fuel oil.

Table III shows results obtained in the analysis of "Bayer liquor." On each sample three results were obtained, caustic, carbonate, and alumina. The method of analysis has been reported by Bell and Marstiller [5] and Watts and Utley [7]. All three titrations were performed with 0.5 N hydrochloric acid to a pH of 8.1. Thirty samples were used and for caustic the standard deviation was 0.35 g/liter. For carbonate the standard deviation was 0.93 g/liter. For alumina the standard deviation was less than 0.1 g/liter.

CONCLUSION

Although all of the work reported here has involved either pH or electrometric titrations, much work has been done on coulometric generation of reagents and colorimetric determination of end points. This work is still in progress. Thus far none of the industrial problems encountered have required these techniques.

Much time and effort has been devoted to the automation of chemical analysis but, the surface has barely been scratched. New principles and approaches to analytical chemistry are required but until they are developed, the best possible use of presently available methods must be made.

REFERENCES

1. Lingane, J. J., Anal. Chem., Vol. 20, p. 285, 1948.
2. Harrison, W. J., Pittsburgh Conference on Analytical Chemistry and Applied Spectroscopy, 1961.
3. Harrison, W. J., Proceedings of the I.S.A., New York City, Fall, 1960.
4. Harrison, W. J., Pittsburgh Conference on Analytical Chemistry and Applied Spectroscopy, 1960.
5. Bell, G. F., and Marstiller, C. M. Ibid.
6. Young, M. G., Plant, A. F., and Harrison, W. J., Proceedings of the I.S.A., Houston, April, 1961 [in print].
7. Watts, H. L., and Utley, D. W., Anal. Chem., Vol. 28, p. 1731, 1956.

THE APPLICATION OF COULOMETRIC BROMINATION TECHNIQUES TO THE DETERMINATION OF UNSATURATED ORGANIC COMPOUNDS

Charles B. Roberts
*The Dow Chemical Company,
Midland, Michigan*

ABSTRACT

A number of unsaturated organic compounds have been determined by coulometric bromination. This method has been used successfully to titrate such compounds as styrene, monochlorostyrene, allyl alcohol, vinyltoluene, divinylbenzene, N-vinylpyrrolidone, vinylbenzylmethyl-ammonium chloride, and other similar compounds. It can be used to determine micro-amounts of such compounds, but has proved equally useful for the determination of assay amounts. It has also been used to determine bromine numbers and iodine numbers of petroleum and natural products.

New instrumental techniques have been developed simultaneously with new applications to make possible more rapid and convenient methods of analysis. A commercial constant-current source covering a current range of 4.82 to 193.0 ma is used in conjunction with an automatic potentiometric end-point detection device designed specifically for this purpose. It does not require a predetermined end-point potential and requires only two preliminary adjustments prior to titration. A redesigned cathode chamber provides a convenient way of emptying and filling the chamber with fresh electrolyte.

Coulometric bromination is a sensitive, accurate, and convenient method of analysis for unsaturated organic compounds. It is more sensitive and selective than the conventional bromination methods, and, through automation, provides a convenience and saving of time valuable in laboratory use.

INTRODUCTION

The first coulometric titration is credited to Szebelledy and Somogyi, who, in 1938, used electrolysis to generate bromine which reacted stoichiometrically with the substance to be determined. The quantity of substance which reacted was computed by Faraday's law from the number of coulombs of electricity rather than from a volume of standard reagent. By keeping the current constant and measuring the time, the number of coulombs was easily calculated.

Many different titrants have since been generated electrolytically, yet bromine still remains the most important for the determination of organic compounds. The number and type of titrants have greatly increased in the past few years, while corresponding improvements in instrument design and operation have also helped to simplify the method and make it more convenient.

Figure 1 shows the instrument that is being used in the laboratories of The Dow Chemical Company in Midland. It consists of three major parts: (1) the source to supply a constant current, (2) a titration cell, and (3) an automatic end-point detection device designed specifically for this purpose. A detailed description of the operation of the instrument will be given later, but at present a brief discussion of the principles of operation and of three methods of determining unsaturated compounds seems in order.

PRINCIPLE OF OPERATION

A suitable electrolyte is placed in the electrolysis cell, the proper current is selected, and, after a blank has been run, an aliquot of the material being determined is added to the cell. The titration is started, the bromine generated reacts with the unsaturated compound, and a potentiometric end-point is obtained. Since the magnitude

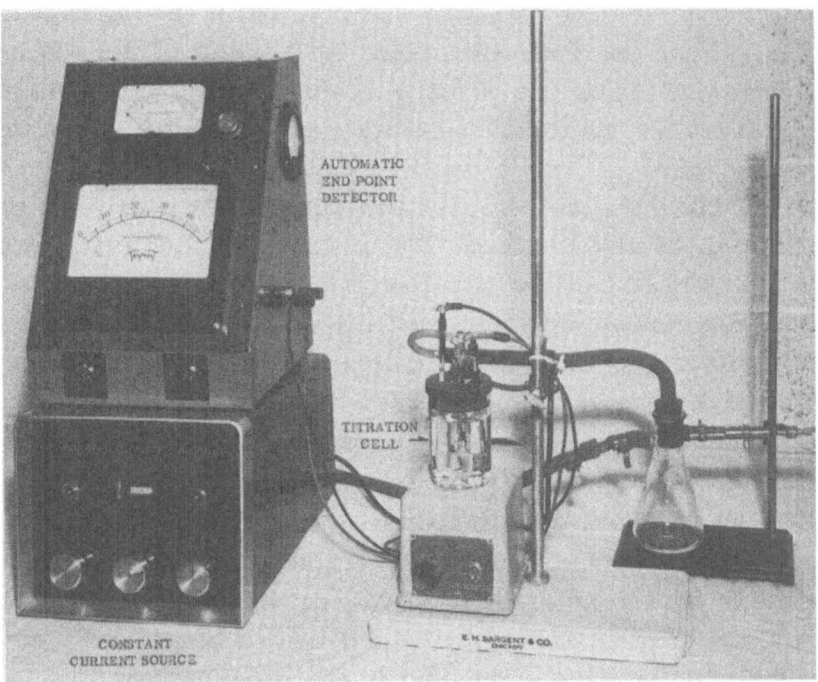

Fig. 1. Automatic coulometric titrator.

of the current is known and the time of its flow can be determined, the number of coulombs can be calculated. From this, the amount of electrolytically generated bromine and the amount of organic compound can be determined.

TITRATION OF ORGANIC COMPOUNDS

Three methods have been developed for titrating organic compounds by coulometric bromination.

The first method is applied to easily brominated compounds. Since the reaction with bromine is rapid, no excess of bromine exists until all of the unsaturated compound has reacted with bromine. Thus, at the end point the voltage will rise rapidly, and if the current is shut off at the inflection point (end point) of the potentiometric curve, the voltage of the solution will remain constant. A recorded potential of this type is shown in Fig. 2.

Table I shows some of the compounds which can be titrated by this method. The column on the left shows those which can be titrated in a 60% methanol—water solution containing sodium bromide and acid. The center column shows compounds which are water soluble and do not require methanol to keep them in solution.

The column on the right shows other organic compounds which will readily brominate even though some are not unsaturated compounds.

The compounds shown in Table II require a slight excess of bromine to force the reaction to completion. After a bromination time which can be empirically determined for each compound, a known amount of arsenic (III) solution is added to the cell. This reacts rapidly

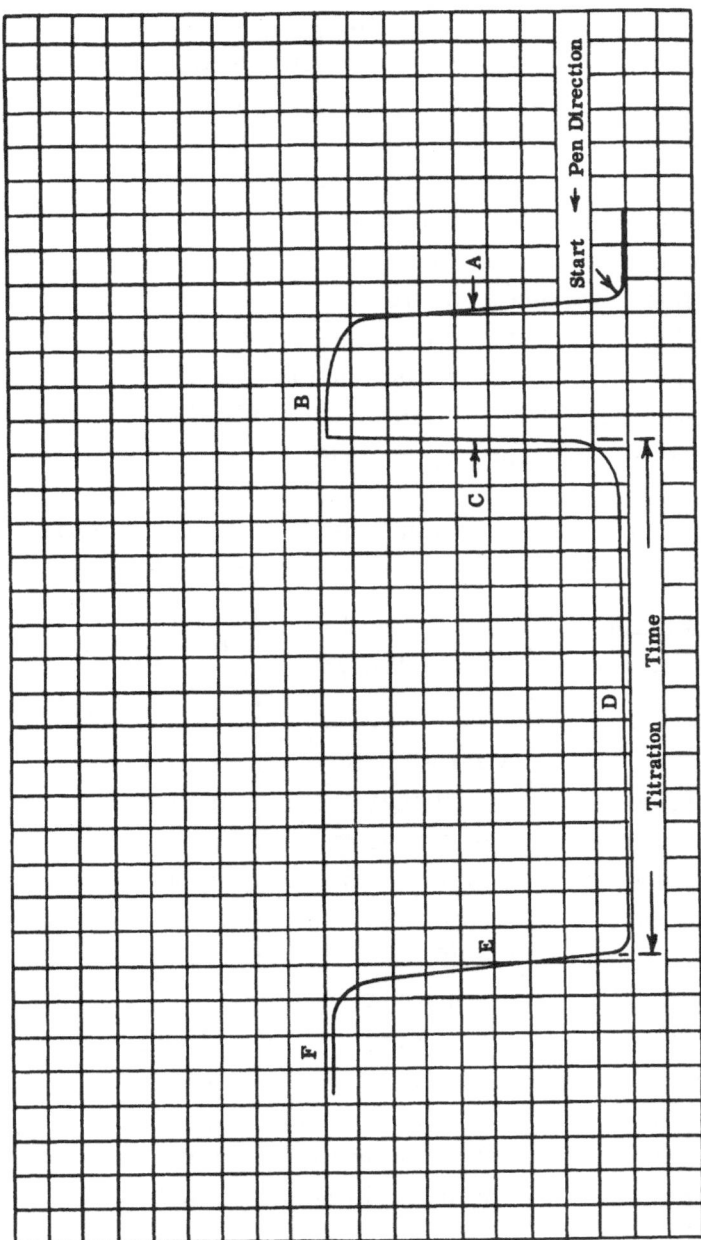

Fig. 2. End-point detection with recording potentiometer. A) Blank titration (increment addition); B) end-point potential; C) addition of sample; D) coulometric bromination of sample; E) incremental addition of bromine near end point; F) final end-point potential.

with the excess bromine and the excess arsenite is then
titrated to an end point by the generation of more bromine. By running a blank, the time required to titrate
added arsenic can be determined. The net time is found
by subtracting the blank time from the total time. The
arsenite strength varies from 0.005 N to 0.02 N depending on the value of current used. The reaction must take
place in an acid medium.

A third method consists of a combination of the
conventional bromate—bromide method and coulometric
bromination. The three compounds shown (Table II) require considerable bromine to complete the reaction and
this is best furnished by an acidified bromate—bromide

Table I. Compounds Which Can Be Brominated Directly

60% methanol—water electrolyte	Aqueous .electrolyte	Other readily brominated compounds
Styrene	Allyl alcohol	Acetanilide
Vinyltoluene	Vinylbenzyltri-	Aniline
Ethylvinylbenzene	methylammo-	Methionine
Divinylbenzene	nium chloride	Thiophene
Alpha methyl-styrene	Sodium styrene sulfonate	Bromine numbers of diisopropyl-
Beta methyl-styrene	n-Vinylpyr-rolidone	benzene, hex-anes, other
Indene	n-Vinyl 3-morpholinone	petroleum products
Vinyl ethers	n-Vinyl 5-methyl-2 oxazolidinone	Iodine number of soybean oil
Vinyl esters (lower mol. wt.)		

solution. It requires a 0.1 N bromate—bromide solution and an arsenic (III) solution which has a slightly greater normality. Thus, if 25 ml of arsenite is added to the bromine produced by 25 ml of the bromate—bromide solution, a slight amount of arsenic (III) is left over. The solution thus serves as its own electrolyte since it contains bromide ion and hydrochloric acid; it is diluted to a suitable volume and the small amount of excess arsenite is titrated with electrolytically generated bromine. This blank should not take over 10 sec at the highest current-generation rate.

The procedure for the sample is similar. An aliquot is placed in a 250-ml iodine flask and 25 ml of the bromate—bromide solution is added. The solution is acidified with hydrochloric acid and allowed to remain in the flask for the desired period of time. Twenty-five milliliters

Table II. Compounds Which Require a Slight Excess of Bromine; Back-Titration with Arsenic (III) Solution

---

Alphamethylstyrene dimers
Styrene dimers
Phenols
Orthophenylphenol
Vinylesters (higher mol. wt.)
Monochlorostyrene
Dichlorostyrene

Modified bromate—bromide—coulometric method

Acrylic acid
Acrylamide
Itaconic acid

---

of the arsenite solution is added, the resulting solution is poured into a titration cell, and the excess arsenic (III) is titrated to an end point. The difference between this time and the blank time represents the amount of bromine consumed by the sample.

The question may arise that since this is bromate—bromide, why not continue in the usual manner. The answer is that coulometric modification increases the accuracy when small amounts of the compound are being determined. Since this method requires a much smaller sample, it can be used whenever a turbidity or color problem occurs with the conventional bromate—bromide method.

INSTRUMENTATION

Within recent years several commercial constant-current sources have appeared on the market; one of these is the Sargent Coulometric Current Source, capable of generating a constant current variable in steps from 4.825 to 193.0 ma. There was then a need for an automatic detection device which could be used with a commercially available source such as that manufactured by Sargent. The detection device shown in Fig. 1 will operate equally as well for slow reaction rates as for fast reaction rates, and has been found to be especially useful for the determination of unsaturated compounds by means of electrolytically generated bromine.

The end-point detector is essentially a low impedance input vacuum-tube voltmeter with a meter relay in series with the indicating meter. The meter relay is used to terminate the generating current at or slightly before the end point, while the indicating meter registers the relative solution potential at all times during the titration. The two meters are synchronized to provide the same reading, but the relay meter no longer functions as a potential-indicating device after the relay is activated; thus, the second or indicating meter indicates the exact end point of the titration. Two adjustments are required; a "zero set" is used to adjust the meters to zero before any titrant has been generated, while a second control is used to adjust the monitoring meter to full scale after an excess of titrant has been generated. After these adjustments have been made, the end-point potential is assumed to be the midpoint as shown on the indicating meter. In practice, the meter relay needle can be set at a current below the midpoint to allow for any overshoot. If, after the generation has been automatically terminated, the indicating meter has not quite reached the midpoint, a few short increments of titrant can be added by means of the "manual" control on the current source until the correct potential is reached. If desired, a few preliminary end points can usually determine the exact setting of the meter relay to provide an accurate, fully automatic end point.

The instrument has been designed to provide for slow reactions where the solution potential may rise prematurely and shut off the titration. As the titrant is consumed upon standing, the solution potential drops and current operation will begin to provide additional titrant. This will continue until a constant potential is established. A delay circuit has been incorporated into the instrument so that a few seconds must elapse between the time the meter relay becomes reactivated and the current is turned on; this prevents a rapid turning off and on of the instrument and makes possible a much smoother end point.

The cathode chamber has been designed for easy removal and addition of the catholyte (Fig. 5). The chamber is made from a medium porosity frit by cutting off one end of a No. 10M sealing tube next to the frit. In order to remove the catholyte, a piece of capillary tubing is inserted to within $\frac{1}{16}$ in. of the frit and sealed into the tube. The cathode is a small piece of platinum foil fastened to a platinum wire and fitted around the capillary tube $\frac{1}{2}$ in. from the frit. The platinum wire is attached to about 4 ft of lead wire and sealed into the top of the tube; a small hole is made in the top of the chamber to allow air to escape, while a small funnel arm at the side facilitates catholyte addition. The portion of the capillary tube protruding from the chamber is bent at right angles

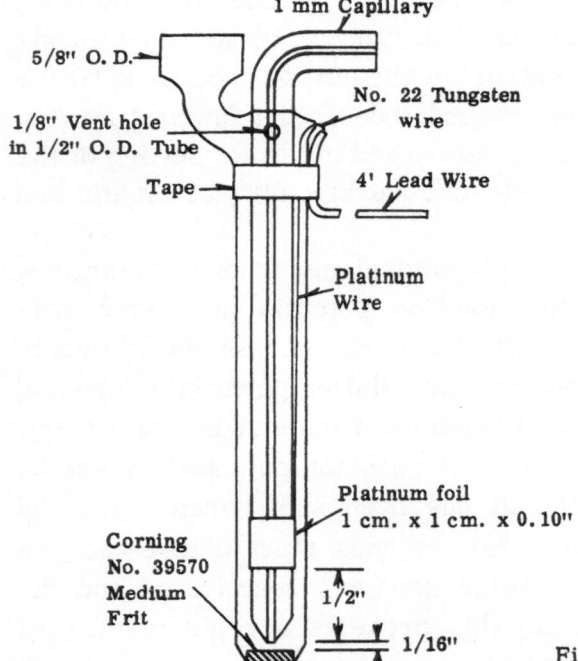

Fig. 3. Cathode assembly.

and connected with plastic tubing to a short, straight piece of glass tubing, which in turn is inserted into a No. 5 stopper. The latter is placed firmly in place in the mouth of a 250-ml filter flask. A filter-bell slide-valve is placed between the side arm of the filter flask and a source of vacuum to allow the chamber to be emptied; this is done by opening the valve so that the vacuum is applied to the filter flask and thus to the cathode chamber through the capillary tubing. The catholyte is immediately removed and collected in the filter flask. The vacuum is shut off and fresh electrolyte is added to the chamber through the small funnel at the top.

SUMMARY

Numerous unsaturated organic compounds can be quickly and accurately determined by coulometric bromination. Three different techniques can be used, depending upon the rate of reaction of the compound with bromine.

An automatic end-point detection device and a convenient cathode chamber have added to the usefulness of a commercially available constant-current source.

COMBINED POTENTIOMETRIC DETERMINATIONS OF MANGANESE, CHROMIUM, AND VANADIUM IN STEELS*

P. A. Bellemare and R. G. Sabourin
Mineral Sciences Division, Mines Branch, Department of Mines and Technical Surveys, Ottawa, Canada.

ABSTRACT

The steel is dissolved in "Spekker Acid." Manganese, chromium, and vanadium are oxidized to Mn^{7+}, Cr^{6+}, and V^{5+}, respectively, by ammonium persulphate in the presence of silver nitrate. Manganese is titrated with sodium nitrite, the equivalence point being indicated at 890 mv. Chromium and vanadium are then titrated together with ferrous ammonium sulphate, the end point occurring at 500 mv. Vanadium is selectively re-oxidized at room temperature by potassium permanganate, and the excess is destroyed with sodium nitrite. The vanadium is then titrated with ferrous ammonium sulfate to the end point at 500 mv, as before.

The method is rapid and accurate. It has the advantage that all three determinations are performed on the one solution.

INTRODUCTION

The most satisfactory chemical methods for estimating Mn, Cr, and V in ferrous alloys are volumetric

*Published with the permission of the Director, Mines Branch, Department of Mines and Technical Surveys, Ottawa, Canada. Her Majesty the Queen, in Right of Canada, reserves the right to reprint this article.

or colorimetric. The advent of automatic titrators offers new scope in the field of physicochemical analysis. The successful combined potentiometric titration of Mn, Cr, and V from a single weighing, without loss of accuracy as compared with previous methods, would result in a considerable saving in time and chemicals.

Theory

A system composed of (a) a solution containing a relatively high concentration of oxidized or reduced ions such as permanganate, dichromate, or ferrous ions, (b) a suitable indicator electrode, and (c) a reference electrode of constant potential, has a given electrode potential which can be read on a suitable potentiometer.

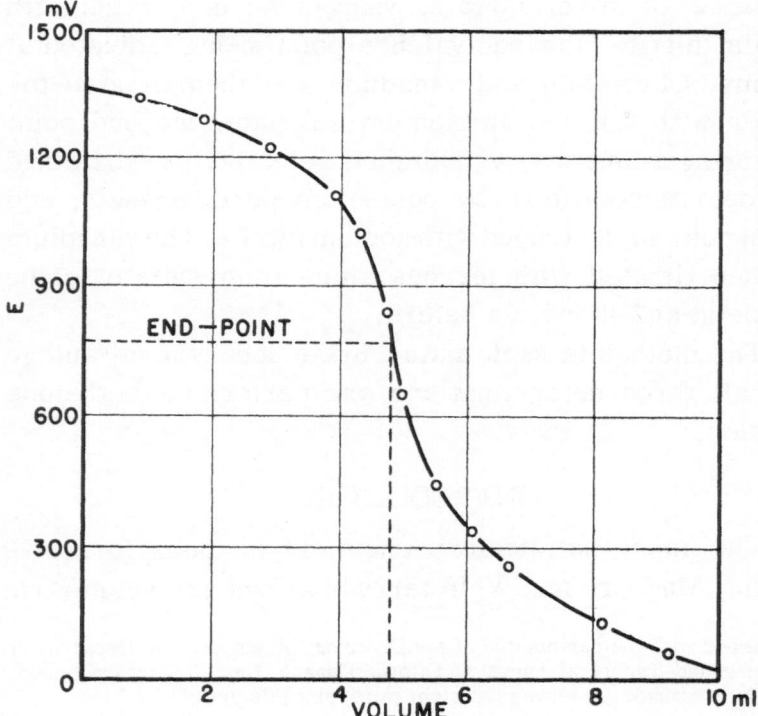

Fig. 1. Titration curve.

As the titration begins, the change in potential is quite small, since the relative change of concentration of oxidized or reduced ions is only slightly affected by a given amount of titrant solution added. Near the end point, however, this relative change of concentration increases rapidly and there is a corresponding rapid change of electrode potential, provided the oxidation and reduction reactions go to virtual completion.

A titration curve is drawn showing the variation of indicator electrode potential with the quantity of titrant solution added, such a curve having the general form illustrated in Fig. 1. Here the sharp change of potential is evident.

In order to determine the end point, one must find the point at which the slope of the titration curve is

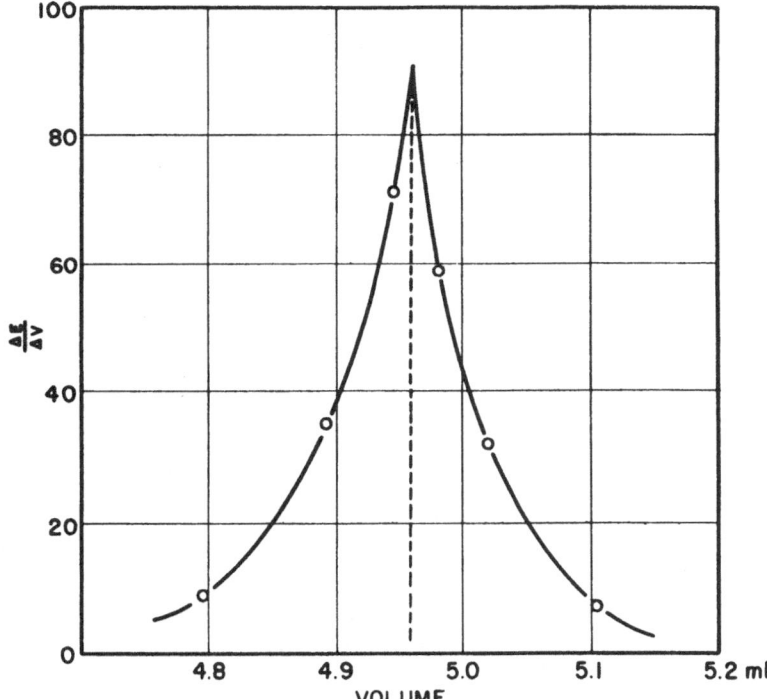

Fig. 2. End-point determination.

maximum. To do so, the general method adopted is to plot $\Delta E/\Delta V$ against the total volume, V, of added titrant (Fig. 2). Provided ΔV is small, $\Delta E/\Delta V$ is a close approximation to the slope of the titration curve and it has a maximum at the end point.

Apparatus

The potentiometer employed in this investigation was a Radiometer Titrator TTT1.* A noncorroding electrode, such as Pt, is used as the indicator electrode. A calomel electrode is used as the reference electrode.

EXPERIMENTAL WORK AND RESULTS

A. Potentiometric Determinations of Mn, Cr, and V in Synthetic Solutions

   1. Potentiometric Titration of Mn in $KMnO_4$, Using Sodium Nitrite as Titrant

     (a) Reagents:

       Potassium Permanganate (stock solution): Dissolve 6.4 g of $KMnO_4$ in 2 liters of distilled water. Boil for 10 to 15 min and let stand overnight. Filter through asbestos on top of glass wool. Standardize against sodium oxalate.

       Sodium Nitrite (0.033 N): Dissolve 13.80 g of sodium nitrite in sufficient H_2O. Dilute to 6 liters.

       "Spekker Acid": 1500 ml of H_2O, 300 ml of H_2SO_4, and 300 ml of H_3PO_4. Add sulfuric

*Manufactured by Radiometer, 72 Emdrupvej, Copenhagen NV, Denmark, and obtained via Canadian Laboratory Supplies Limited, Box 2090, Station O, Montreal 9, Quebec.

acid to the H_2O with constant stirring. Cool, then add H_3PO_4.

(b) Procedure:

Each synthetic solution contains a known concentration of $KMnO_4$. To all synthetic solutions employed for the following titrations, add 25 ml of "Spekker Acid" and bring the volume up to 150 ml with distilled H_2O.

Table I shows the results obtained when Mn^{7+} was titrated with sodium nitrite (0.033 N) in order to find the experimental end point.

2. Potentiometric Titration of Cr in $K_2Cr_2O_7$, Using Ferrous Ammonium Sulfate as Titrant

(a) Reagents:

Potassium Dichromate (stock solution): Weigh 9.806 g of dry potassium dichromate. Dissolve in sufficient distilled water and dilute to 2 liters.

Table I. End-Point Determination. Titration of Mn^{7+} with $NaNO_2$ (0.033 N)

$KMnO_4$ solution No.	Mn, %	Experimental end point, mv
1	0.018	890
2	0.089	845
3	0.177	920
4	0.885	880
5	1.77	887
6	2.65	900

Ferrous Ammonium Sulfate (0.2 N): 156 g of ferrous ammonium sulfate. 100 ml of H_2SO_4. Dissolve the ferrous ammonium sulfate in sufficient water, add H_2SO_4, cool, and make up to 2 liters.

"Spekker Acid": e.g., as above.

(b) Procedure:

Each synthetic solution contains a known concentration of $K_2Cr_2O_7$. To all synthetic solutions to be titrated, add 25 ml of "Spekker Acid" and dilute to 150 ml.

Table II shows the results obtained from the titration of Cr^{6+} with ferrous ammonium sulfate (0.2 N) in order to establish the experimental end point.

3. Potentiometric Titration of V in V_2O_5, Using Ferrous Ammonium Sulfate as Titrant

(a) Reagents:

Vanadium Pentoxide: Dissolve 0.8 g of V_2O_5 in

Table II. End-Point Determination. Titration of Cr^{6+} with Ferrous Ammonium Sulfate (0.2 N)

$K_2Cr_2O_7$ solution No.	Cr, %	Experimental end point, mv
1	0.018	600
2	0.046	604
3	0.092	600
4	0.184	630
5	3.68	600
6	5.52	600

100 ml of $1:3$ H_2SO_4. Dilute to 500 ml with distilled water.

Ferrous Ammonium Sulfate (0.02 N).

"Spekker Acid": e.g., as above.

(b) Procedure:

Each synthetic solution contains a known concentration of V_2O_5. To all synthetic solutions to be titrated add 25 ml of "Spekker Acid" and dilute to 150 ml.

Table III shows the results obtained from the titration of V^{5+} with ferrous ammonium sulfate (0.02 N) in order to establish the experimental end point.

4. Combined Determinations of Mn, Cr, and V in Synthetic Solutions

Procedure:

Each solution titrated contains various concentrations of Mn, Cr, and V. To all solutions to be titrated add 25 ml of "Spekker Acid" and dilute to 150 ml. The data in Figs. 3, 4, and 5 were

Table III. End-Point Determination. Titration of V^{5+} with Ferrous Ammonium Sulfate (0.02 N)

V_2O_5 solution No.	V, %	Experimental end point, mv
1	0.18	510
2	0.45	500
3	0.90	500
4	1.35	490
5	1.80	510

obtained from the titration of Mn, Cr, and V
performed on the same solution.

5. Discussion

The various experimental end points obtained for
the determination of the equivalence point (Tables
I, II, and III) indicate some difficulty in obtaining
constant values. The arithmetic mean values for
Mn, Cr, and V end points are 887, 606, and 502
mv, respectively. However, for practical purposes the equivalence point for Mn is established
at 890 mv, for Cr (V absent) at 600 mv, for
Cr + V at 500 mv, and for V at 500 mv.

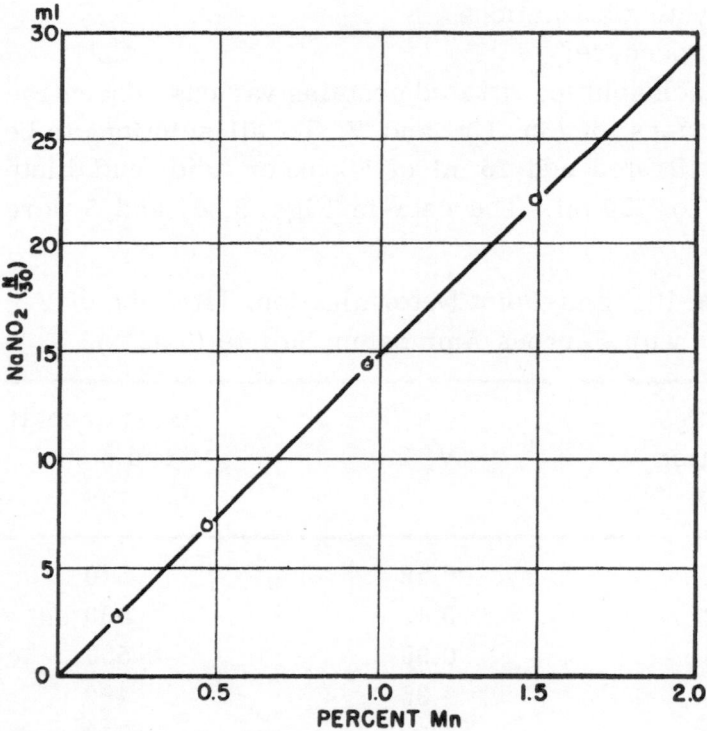

Fig. 3. Titration of Mn^{7+}.

The linearity of the curves of Figs. 3, 4, and 5 attests to the reliability of the method at various concentrations. Since a blank* is included in the titration of Cr and V, the curves in Figs. 4 and 5 intersect the ordinate above the zero. There is no evidence of interference in the titration of a solution containing Mn, Cr, and V. The results obtained are reproducible to ±0.01% Mn, ±0.02% Cr, and ±0.005% V.

*The Cr blank can be determined by back-titration with $KMnO_4$ after reduction of Cr + V. The V blank can also be determined by back-titration with $KMnO_4$ after the reduction of V.

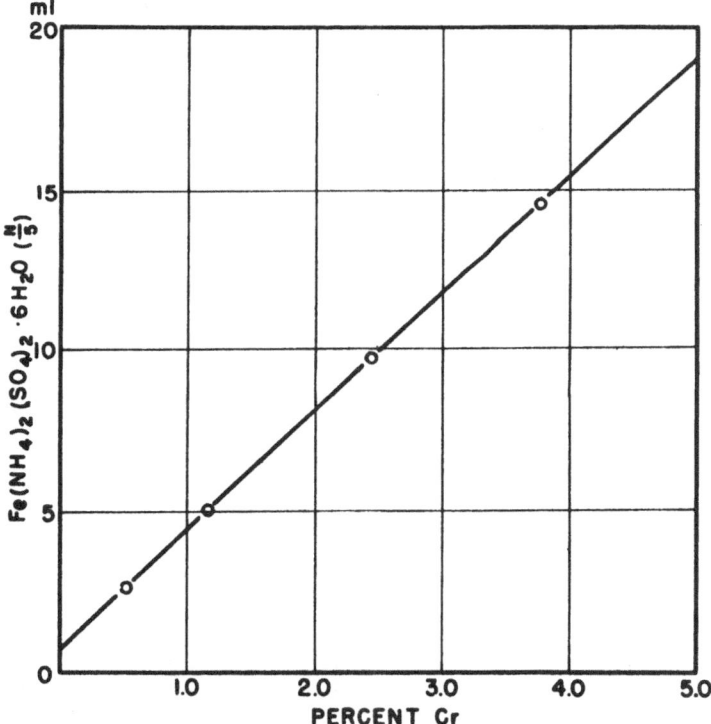

Fig. 4. Titration of Cr^{6+}.

B. The Combined Determinations of Mn, Cr, and V in
    Steels
    1. Analytical Procedure
       (a) Reagents:
           HNO_3 (conc.)
           $KMnO_4$ solution
           "Spekker Acid"
           Ferrous ammonium sulfate (0.2 N)
           Ferrous ammonium sulfate (0.02 N)
           Sodium nitrite (0.033 N)
           Analoids*: Silver nitrate No. 623
                      Ammonium persulfate No. 8b
                      Urea No. 31

*Ridsdale and Co., Ltd., Newham Hall, Middlesborough, England.

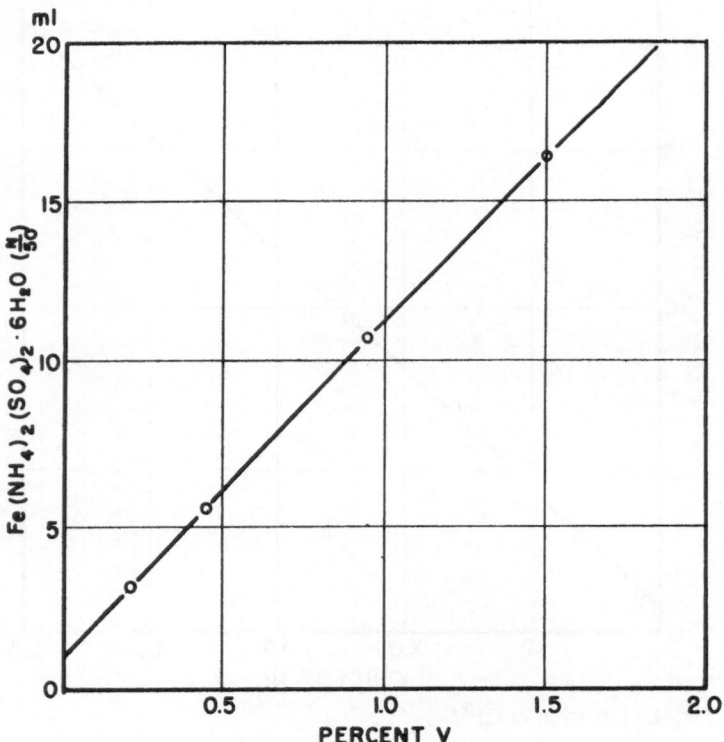

Fig. 5. Titration of V^{5+}.

(b) Method:

Weigh 1.0 g of sample into a 400-ml beaker. Add 40 ml of "Spekker Acid."*

Simmer until all but the black insoluble carbides have dissolved. Add HNO_3 dropwise until oxidation is complete, and simmer for a few minutes. Cool, then dilute to 150 ml with H_2O. Add a few glass beads and bring to a boil. Add one analoid of silver nitrate No. 623 and dissolve. Add one to three analoids of ammonium persulfate No. 8b and bring to a boil. Heat until the Mn, Cr, and V are completely oxidized and the excess persulfate is destroyed. Cool slightly and dilute to 225 ml. Titrate Mn with $NaNO_2$ (0.033 N), setting the end point on the titrator at 890 mv. When the end point is reached, add one analoid of urea No. 31 and stir for a few minutes on a magnetic stirrer. Titrate for Cr + V with ferrous ammonium sulfate (0.2 N), using an end-point setting of 500 mv. Cool to between 25 and 30°C. Re-oxidize V with $KMnO_4$ solution, adding a few drops in excess to ensure complete oxidation of V; then, while stirring, slowly drop in sodium nitrite to completely discharge the permanganate color, and add a few drops in excess. Add 0.7 g of urea analoid No. 31. Stir until reaction ceases and allow to stand a few minutes. Titrate V with ferrous ammonium sulfate (0.02 N), using an end-point setting of 500 mv.

*Steels not readily soluble in "Spekker Acid" may be dissolved in aqua regia. "Spekker Acid" is then added and evaporated to fumes of H_2SO_4 to remove all chlorides.

Standard curves are established using National Bureau of Standards samples; these curves are in agreement with those drawn in Figs. 3, 4, and 5, obtained from synthetic solutions.

2. Conclusion

This technique is applicable to a wide range of manganese, chromium, and vanadium found in steels. At the lower percentage levels, titrant solutions of weaker concentrations are used.

Inasmuch as all three elements can be determined rapidly and accurately on the one solution, the combined method presented in this paper is an improvement over previous volumetric and colorimetric methods.

REFERENCES

1. Furman, N. H. Anal. Chem. Vol. 22, p. 33, 1950.
2. "Analoid Methods of Analysis," Folder No. 340 R (reprinted Nov., 1958) issued by Ridsdale and Co., Ltd., Newham Hall, Middlesborough, England, pp. 61-70.
3. Glasstone, S. The Elements of Physical Chemistry (D. Van Nostrand Co., Inc., New York, 1954), 10th ed., pp. 455-457.

Complexometric Titrations

THE USE OF ETHYLENEDIAMINE TARTRATE IN POLAROGRAPHY

I. The Determination of Iron, Nickel, Cadmium, and Copper*

Thomas Bruce and R. W. Ashley
Atomic Energy of Canada Limited
Chalk River, Ontario

ABSTRACT

Yoshiaki Miura has published [1] a procedure for the determination of iron and copper in aluminum alloys, using ethylenediamine tartrate (EDT) and potassium pyrophosphate as base electrolyte. As a result of this investigation, polarographic procedures using ethylenediamine tartrate complexes have been developed for the determination of iron (ferrous), nickel, cadmium, and copper, as well as for mixtures of the latter three elements.

The essential features of these procedures are given in Table I. It is necessary, in an iron determination, to modify the procedure; the electrolyte solution is made 1% in $NH_2OH \cdot HCl$ and a small volume of chloroform is added to the electrolysis cell to place a barrier between the iron solution and the collected mercury.

Table I shows that (1) nickel and cadmium will not interfere with an iron or copper determination in the

---

*The full text of this paper was not made available for publication.

Table I. Polarography of Metal-Ethylenediamine Tartrate Complexes

Polarographic medium	Metal ion	Concentration range	pH range	$E_{\frac{1}{2}}$ (vs SCE)	$\Delta C / \Delta i_d$
0.5 M Na · EDT, 0.1 M Na$_4$P$_2$O$_7$	Fe^{2+}	1–10 mg/50 ml	5.0–6.0	−0.33 v	1.52 ± 0.03
	Ni			No wave	
	Cd			−0.68 v	
	Cu			−0.3 v	
0.5 M NH$_4$ · EDT, 0.1 M NH$_4$Cl	Fe^{2+}			No wave	
	Ni	0.1–2 mg/50 ml	7.1–7.5	−1.02 v	0.602 ± 0.012
	Cd	0.5–5 mg/50 ml	6.5–7.5	−0.68 v	1.08 ± 0.02
	Cu			−0.3 v	
0.5 M Na · EDT, 0.1 M NaCl	Cu	0.1–2 mg/50 ml	6.8–7.2	−0.3 v	0.572 ± 0.014

NaEDT · $Na_4P_2O_7$ system, but that the latter two elements cannot be determined in the same solution. Nickel is not reduced in this electrolyte. (2) Nickel, cadmium, and copper can be determined together by the $NH_4EDT · NH_4Cl$ procedure. Iron is not reduced in this system. (3) Copper is reduced in a NaEDT · NaCl base electrolyte.

This investigation suggests that ethylenediamine tartrate may be a useful reagent for the polarographic determination of a variety of metal ions. A similarity to ethylenediaminetetraacetic acid complexes, also used in polarography [2], is to be expected.

REFERENCES

1. Miura, Y. Anal. Abstracts, Vol. 6, p. 3427, 1959.
2. Welcher, F. J. "The Analytical Uses of Ethylenediaminetetraacetic Acid" (D. Van Nostrand Co., Inc., Toronto, 1958).

A NEW INSTRUMENT FOR AUTOMATIC COLORIMETRIC AND FLUORIMETRIC TITRATIONS*

H. K. Howerton and J. C. Wasilewski
American Instrument Company, Inc.
Silver Spring, Maryland

ABSTRACT

A new instrument for automatic colorimetric and fluorimetric titrations is described. Titrant delivery is automatically stopped at the end point by the use of a microammeter with a built-in relay which stops a motor-driven buret equipped with digital read-out or a gravity driven buret equipped with solenoid control. Monochromatic light is used to excite fluorescence, which is received at 90° by a photomultiplier tube with the microammeter in its anode circuit. Light transmitted through the titration vessel is detected by a barrier-layer photocell, the output of which is fed to the microammeter.

Results are given for

(1) automatic acid-base (HCl-NH₄OH) titration using a mixed colorimetric indicator (brom cresol green-methyl red),

(2) automatic determination of zinc or standardization of EDTA (ethylenediaminetetraacetic acid) using Superchrome Black TS as the colorimetric indicator, and

*Parts of this paper were presented at the Pittsburgh Conference on Analytical Chemistry and Applied Spectroscopy, Pittsburgh, Pennsylvania, February 27, 1961.

(3) automatic determination of calcium based on complexometric titration using EDTA with Calcein as the indicator of fluorescence end point.

Automatic data presentation is achieved by printing equivalence volume in microliters or by recording complete titration curves on an X—Y recorder.

Precision, accuracy, and range of the method are discussed.

TITRACOLORMAT

Titrocolormat is the name we have given to the basic titration instrument which consists of a metal cabinet housing electrical components and supporting optical components, a magnetic stirrer, and a titration vessel. A support rod is provided for burets and refill vessels. Electrical outlets in the rear of the cabinet provide for connection to power, lamp, buret, stirrer, photocell, recorder, and printer. The titration vessels are held on top of a magnetic stirrer which is raised and lowered by a lever control, placing the vessels into or out of the optical beam.

Optical System

The optical system is shown in Fig. 1. It consists of a tungsten lamp,* a lens to collimate light from the lamp through the titration vessel and onto the barrier-layer photocell,† a lens to image fluorescent light from the titrate‡ onto the photomultiplier tube,§ and filters to make the light from the lamp and titrate monochromatic.

*General Electric Company, Schenectady, New York, No. 1323.
†International Rectifier Corporation, El Segundo, California, No. A5-M.
‡The titrate is the substance titrated.
§Radio Corporation of America, Lancaster, Pennsylvania, Type 1P21.

An adjustable slit and on-off shutter control the light intensity reaching the photomultiplier tube.

Electrical System

The electrical system consists of a constant-voltage transformer supplying controlled power to the lamp and a D.C. power supply for the microammeter relay reset functions. A 900-v battery, consisting of three 300-v batteries, is used to supply the photomultiplier tube dynode voltage and the anode is connected directly to the microammeter for indication of fluorescent intensity. The titration is stopped automatically by the micro-ammeter (and its built-in relay) when the signal from the photocell (or photomultiplier tube) reaches a value determined by the control pointer.

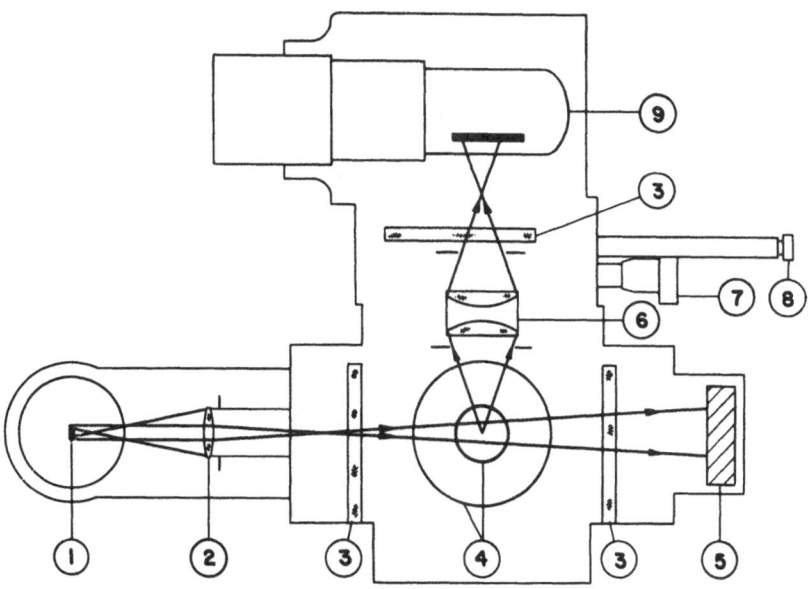

Fig. 1. Optical diagram. 1) Tungsten source; 2) lens; 3) filters; 4) titration vessel; 5) photocell; 6) lens assembly; 7) adjustable slit; 8) on—off shutter; 9) 1P21 photomultiplier tube.

Titration Vessels

The following six types of vessels can be accommodated: three standard beakers of 20-, 30-, and 100-ml capacity; two vials, approximately $3/4$ in. in diameter and $1\frac{1}{2}$ in. high (9 ml capacity) and $1\frac{1}{4}$ in. in diameter and 2 in. high (15 ml capacity); and a 10 by 10 by 48-mm-high cuvette. The latter is convenient in performing spectrophotometric or spectrophotofluorometric measurements in other instruments in order to follow optical density and fluorescence changes.

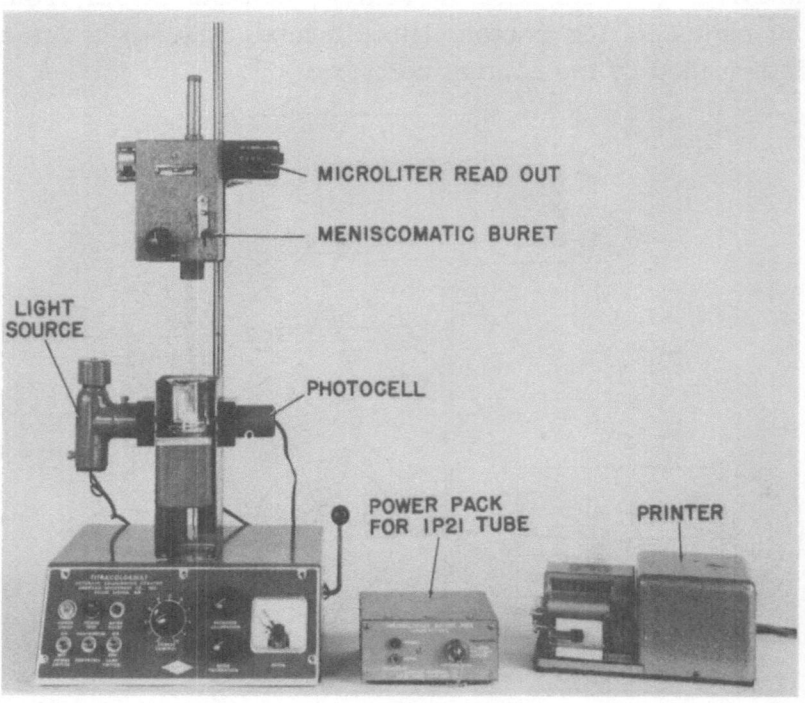

Fig. 2. Titracolormat equipped with 3.5-ml-capacity Meniscomatic buret and printer.

BURETS

We used a Meniscomatic buret [1] whose capacity was approximately 3.5 ml (Fig. 2) and one whose capacity was approximately 35 ml (Fig. 3). These burets have been described by Klaasse [2]. They are motor-driven and equipped with Vycor* plungers. Digital display of titrant volume is provided by a five-digit counter, the last digit corresponding to 0.0001 ml for the 3.5-ml-capacity buret,

*96% silica, Corning Glass Works, Corning, New York.

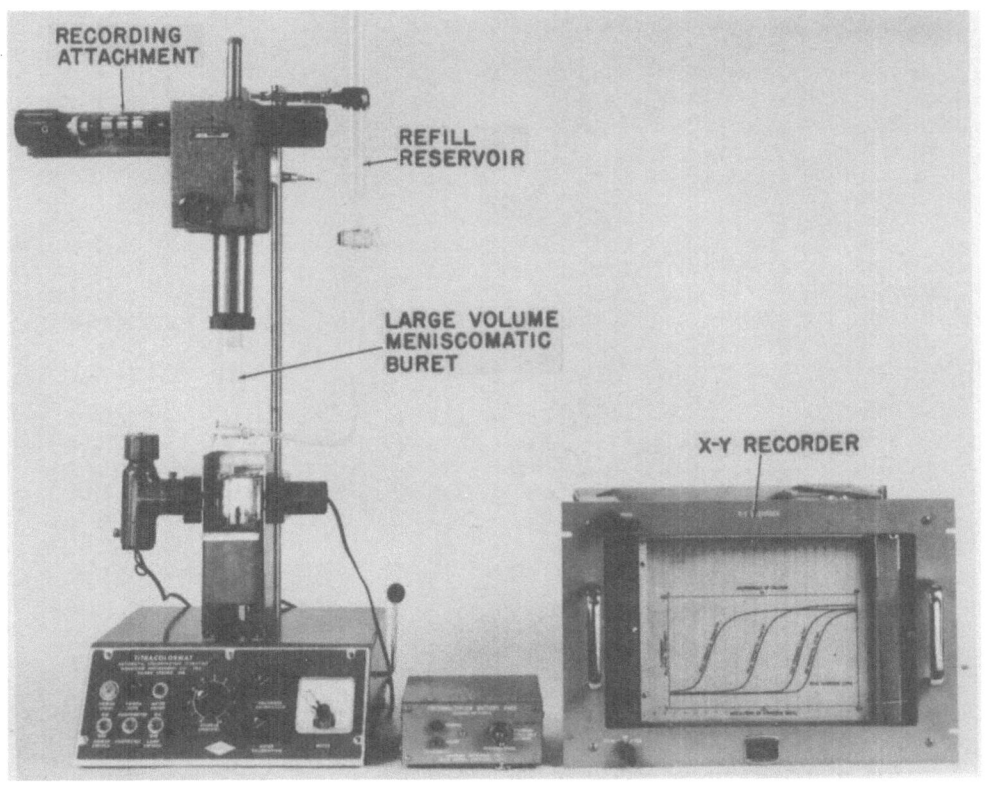

Fig. 3. Titracolormat equipped with 35-ml-capacity Meniscomatic buret, X—Y recorder, and refill reservoir.

and 0.001 ml for the 35-ml-capacity buret. The recording
attachment shown in Fig. 3 was used to translate titrant
volume into an electrical signal applied to the recorder.
This attachment incorporated a fixed-ratio speed changer

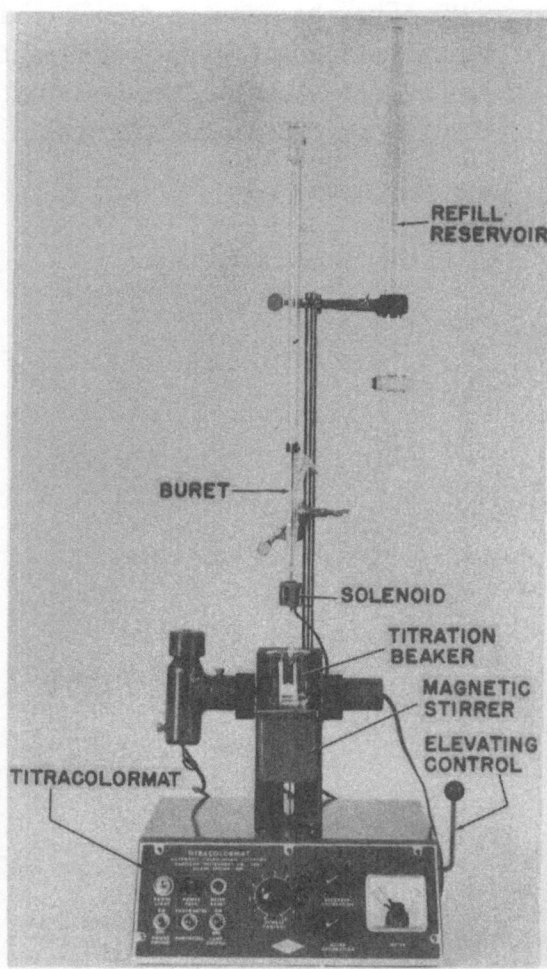

Fig. 4. Titracolormat equipped with solenoid-
controlled buret and refill reservoir.

(1 to 350), a ten-turn helical potentiometer, a variable resistor, and a mercury battery—all mounted on a bracket for attachment to the buret.

In addition, we used a gravity-fed buret* with solenoid control (Fig. 4).

The refill reservoirs shown in Figs. 3 and 4 permitted rapid refilling of the burets.

DATA PRESENTATION

Printer

An electromechanical printer† (Fig. 2) was used for permanent recording of titration volumes. It was energized by an electrical signal from the contacts of a microswitch mounted on the Meniscomatic buret. The microswitch was actuated by a cam attached to the buret counter shaft. At the end of a titration, when the buret was automatically stopped, the titrant volume was printed on a paper strip.

X—Y Recorder

An E.I.‡ Model 101 X—Y recorder was used to record titration curves. The X axis was fed from the buret recording attachment and the Y axis from the photomultiplier tube or photocell output. The pen was slightly modified to accept India ink and the recording was made directly on $8\frac{1}{2}$ by 11 in. linen placed on the recorder bed.

*Houston Glass Fabricating Co., Houston, Texas.
†Presin Co., Santa Monica, California, No. F272.
‡Electro Instruments, Inc., 3540 Aero Court, San Diego, California.

H. K. Howerton and J. C. Wasilewski

COLORIMETRIC TITRATIONS

Acid—Base

Reagents

Titrant: 0.2856 N hydrochloric acid.

Titrate: 0.3476 N ammonium hydroxide.

Buffer solution: 4%, by weight, boric acid (aqueous).

Indicator: Brom cresol green-methyl red solution prepared by mixing 5 parts BCG solution and 1 part MR solution, each made by dissolving 0.1 g of the indicator in 100 ml of ethanol.

Instrumental Parameters

Filter: Cat. No. 33-78-60 interference filter,* 35% transmission at 600 mμ, half-width of 8 mμ, placed between the lamp and titration vessel.

Meniscomatic buret: 3.5 ml capacity.

Titration vessel: $\frac{3}{4}$ in. in diameter, $1\frac{1}{2}$ in. high, approximately 9 ml capacity.

Procedure

Preliminary adjustments: The speed of rotation of the magnet in the titrate was adjusted to provide a vortex just sufficient for proper mixing but not large enough to interfere with the light beam. Rate of titrant delivery was adjusted to approximately 600 μl/min. A constant titrate volume was used by adding water to measured amounts of ammonium hydroxide, buffer, and indicator mixture.

Adjustment of "difference current": With the titrate in the light beam, the microammeter pointer was adjusted to a convenient value, e.g., 5 μa. The

---

*All interference filters used were manufactured by Bausch and Lomb Optical Co., Rochester, New York.

control pointer was then set 2 μa higher to 7 μa. A titration was performed and the titrant volume noted at the automatic cut-off. If the titrant volume was below the stoichiometric value, the control pointer was readjusted to a higher value, e.g., 8 μa, and if above, to a lower value, e.g., 6 μa. Another titration was performed, the titrant volume noted, and the control pointer readjusted accordingly. It was usually possible to obtain an equivalence volume within 1% of stoichiometric after two or three titrations. This "difference current" was used for subsequent titrations.

Results

Table I shows results of repeated titrations with automatic stop using 2-μa difference current for this acid—base system. From the stoichiometry of this neutralization, 1 ml of titrate is equivalent to 0.122 ml of titrant. The accuracy was approximately 3%. The standard deviation from the arithmetical mean, about 0.5%, was calculated using inefficient statistics [3].

Zinc—EDTA

Reagents

Titrant: 0.02 M zinc chloride, prepared by dissolving the appropriate amount of zinc metal* in a minimum of hydrochloric acid and diluting to volume.

Titrate: 0.002 M EDTA.

Buffer: Ammonium chloride (67.5 g) and 570 ml of concentrated ammonium hydroxide were made up

*Fisher Scientific Co., No. Z11, 99.99% pure.

to 1 liter with water. Approximately 1 ml was used
for each titration.

Indicator: 0.2 g of Superchrome Black TS was tri-
turated with 100 g of ACS grade sodium chloride.
0.2 g of this mixture was added to the titration
vessel.

Instrumental Parameters

Filter: Cat. No. 33-78-55 interference filter, 35%
peak transmission at 553 mμ, half band width of
8 mμ, placed between the lamp and titration vessel.

Table I. Repeated Titration (with Automatic
Stop) of NH$_4$OH with HCl

Trial	Milliliters of 0.3476 N NH$_4$OH titrated		
	3.0	2.0	1.0
1	0.3571	0.2360	0.1209
2	0.3575	0.2365	0.1216
3	0.3547	0.2370	0.1201
4	0.3541	0.2361	0.1190
5	0.3526	0.2359	0.1214
6	0.3533	0.2382	0.1195
7	0.3561	0.2357	0.1188
8	0.3533	0.2361	0.1204
9	0.3557	0.2354	0.1192
10	-	0.2364	0.1187
11	-	-	0.1200
12	-	-	0.1187
13	-	-	0.1204
14	-	-	0.1201
Arithmetic mean	0.355 ml	0.236 ml	0.199 ml
Standard deviation, %	0.5	0.5	0.7
Stoichiometric equivalent	0.366 ml	0.244 ml	0.122 ml
Accuracy, %	3	3	3

In all cases 1.0 ml of boric acid buffer solution and 2 drops
of indicator solution were added to measured volumes of
0.3476 N NH$_4$OH and the mixture was diluted with water to a
fixed volume in the titration vessel. The titrant was 0.2856 N
HCl and the entries represent milliliters of this acid added for
automatic stop. The procedure and instrumental parameters are
given in the text.

Meniscomatic buret: **3.5 ml** capacity, equipped with
recording attachment.

Titration vessel: Beaker $2\frac{1}{4}$ in. in diameter, $2\frac{3}{4}$ in.
high, approximately 100 ml capacity.

X—Y Recorder: The X axis was set so that about
0.05 ml of titrant corresponded to 1 in. deflection.
The Y axis was set at 10 mv/in. and was connected
directly across the photocell.

Procedure

A measured amount of EDTA, buffer, and indicator
was added to the titration vessel and the Y axis of
the recorder was adjusted to give approximately
4 in. deflection. The titration was started and allowed to proceed past the end point in order to record
a dynamic titration curve for this system.

Results

Figure 5 shows two titration curves obtained using 1
and 3 ml of EDTA solution. From the stoichiometry
of this chelation, 1 ml of titrate is equivalent to
0.1 ml of titrant. By graphically locating the titrant
volume at the inflection point, an agreement within
1.5% of stoichiometric was obtained. Better accuracy
would result if cross-section paper and an expanded
X axis with zero supression were employed. Although we did not study this system statistically
using automatic stop, there is no reason why it
could not be used to standardize EDTA or to determine zinc.

FLUORIMETRIC TITRATIONS

An ideal fluorimetric titration is one performed in
such a manner that a highly fluorescent product is formed

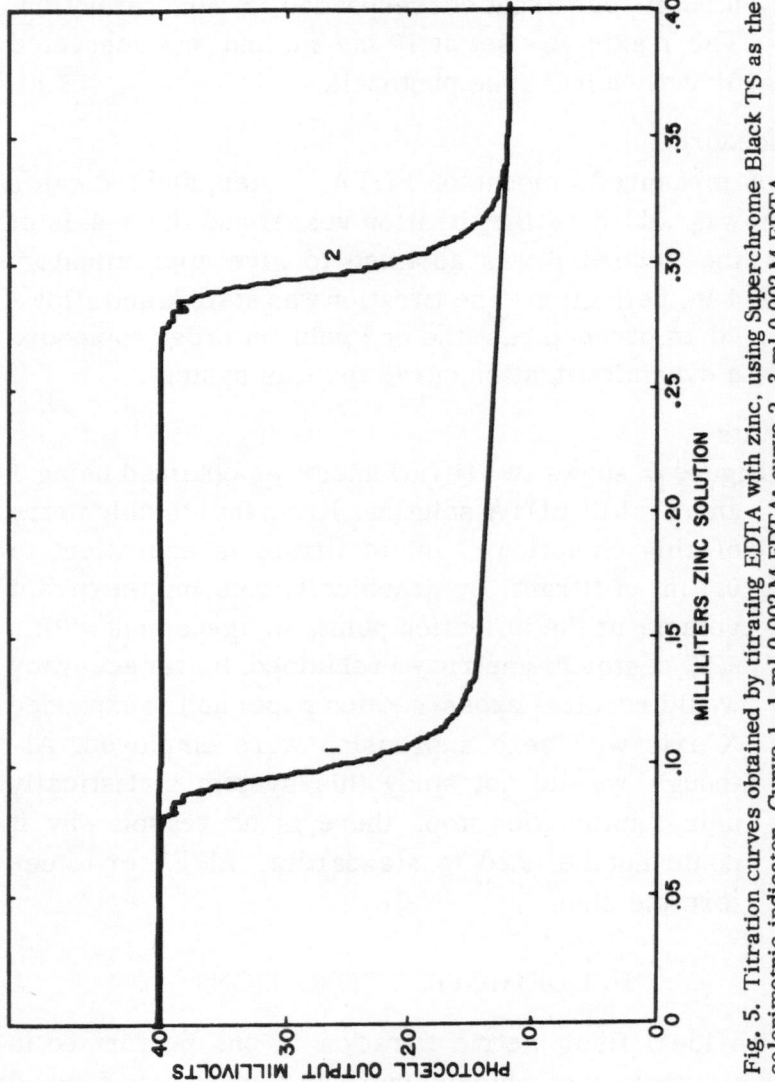

Fig. 5. Titration curves obtained by titrating EDTA with zinc, using Superchrome Black TS as the colorimetric indicator. Curve 1, 1 ml 0.002 M EDTA; curve 2, 3 ml 0.002 M EDTA.

and its fluorescence detected at the equivalence point. Since fluorescence intensity is usually a linear function of concentration, it is difficult to detect the feeble fluorescence appearing at the equivalence point, especially if low concentrations are used. For best results, it is necessary to excite the titrate at or near the wavelength of maximum excitation and detect the fluorescence at or near the wavelength maximum emission. The indicator concentration is important. If it is too low, the fluorescent signal will not be great enough to detect. If it is too high, there will be concentration quenching* and a corresponding reduction in fluorescent signal if 90° geometry is used.

We studied the calcium—EDTA chelation using Calcein [4] as the fluorescent indicator. This indicator forms a fluorescent complex with free calcium ions. In order to determine calcium in a sample, the excess EDTA remaining after stoichiometric chelation of calcium in the sample is "back-titrated" with a known concentration of calcium solution. The equivalence point is reached when all of the excess EDTA is chelated and is detected by fluorescence of the Calcein-calcium complex formed by a slight excess of free calcium ions. The calcium content of the sample (in micrograms) is equal to the volume (in milliliters) of titrant added when no sample was introduced, minus the volume (in milliliters) of titrant added when the sample was introduced, times the calcium content of the titrant (in micrograms per milliliter).

In order to select suitable filters for exciting the titrate maximally and for isolating the fluorescence from scatter, the excitation and emission spectra of

---

*The term "quenching" has often been employed in a vague manner to cover any effect reducing the measurable fluorescence or making it nonlinear with concentration.

Calcein were recorded using a spectrophotofluorometer
[5]. Figure 6 shows the spectra obtained. Since the wavelength for maximum excitation was approximately 470 mμ,
a Cat. No. 33-78-47 interference filter with peak transmission at 470 mμ was selected as the exciting filter.
Similarly, a Cat. No. 33-78-52 interference filter with
peak transmission at 524 mμ was selected as the emission filter.

The optimum Calcein concentration was determined
by titrating 20 ml of 0.0002 M EDTA with 0.00025 M
calcium chloride and the resulting titration curves recorded at four different calcein levels. One-fourth gram
of Calcein was dissolved in 4 ml of 1 N NaOH and 30 ml
of water was added. This solution was then diluted 1 : 100
and 0.1, 0.5, 1, and 2.5 ml of the dilute solution was
added to the EDTA for each titration. The titration curves
are shown in Fig. 7. The curve labeled "0.5 ml Calcein"
is steepest at the equivalence point, consequently this
concentration (approximately 2 μg/ml) was used for
subsequent titrations.

Calcium in Lab-trol and Patho-trol

Lab-trol* and Patho-trol* are control sera containing
known concentrations of various substances. The analysis
furnished for the samples used was: for Lab-trol, total
calcium, 100 μg/ml; magnesium, 32 μg/ml; sodium,
3.33 mg/ml; and potassium of 191 μg/ml; for Patho-trol,
total calcium, 81 μg/ml; magnesium, 54.3 μg/ml; sodium,
2.74 mg/ml; and potassium, 72 μg/ml. The Clark-Collip
[6] method was used in determining the calcium content
of both samples according to the analysis.

*Dade Reagents, Inc., Miami, Florida.

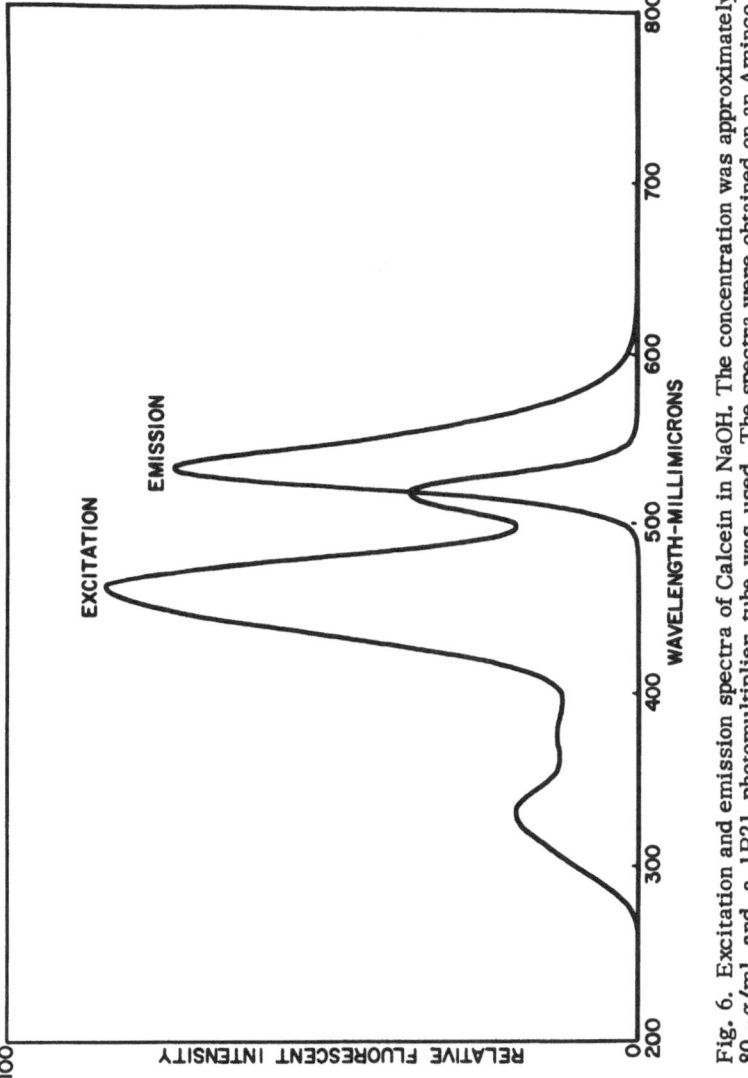

Fig. 6. Excitation and emission spectra of Calcein in NaOH. The concentration was approximately 80 μg/ml and a 1P21 photomultiplier tube was used. The spectra were obtained on an Aminco-Bowman Spectrophotofluorometer equipped with polarizers in the excitation and emission beams to reduce the intensity of scatter peaks, which interfere when the excitation and emission wavelengths are equal.

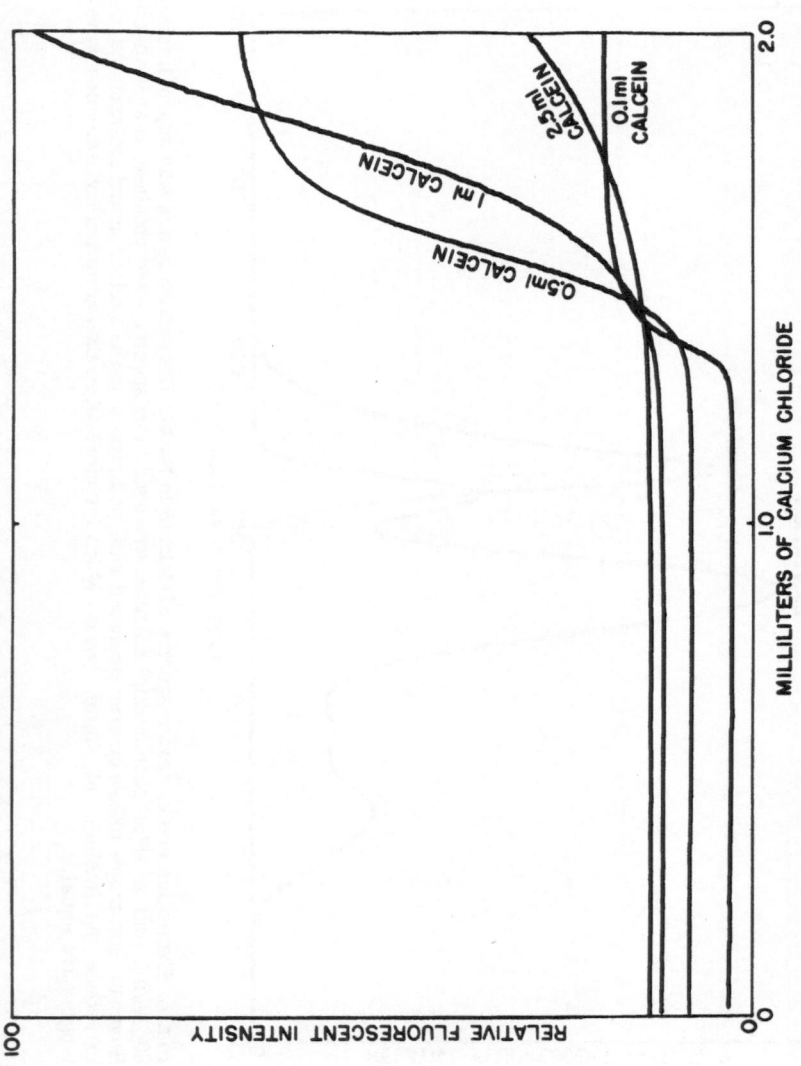

Fig. 7. Titration curves obtained using different concentrations of Calcein. Twenty milliliters of 0.00002 M EDTA was titrated. 0.1, 0.5, 1, and 2.5 ml of Calcein solution (80 μg/ml) was added to the titrate. The steepest curve was obtained using 0.5 ml of Calcein. The signal was weak when using 0.1 ml, due to the low concentration. It was also weak when using 2.5 ml, due to concentration quenching.

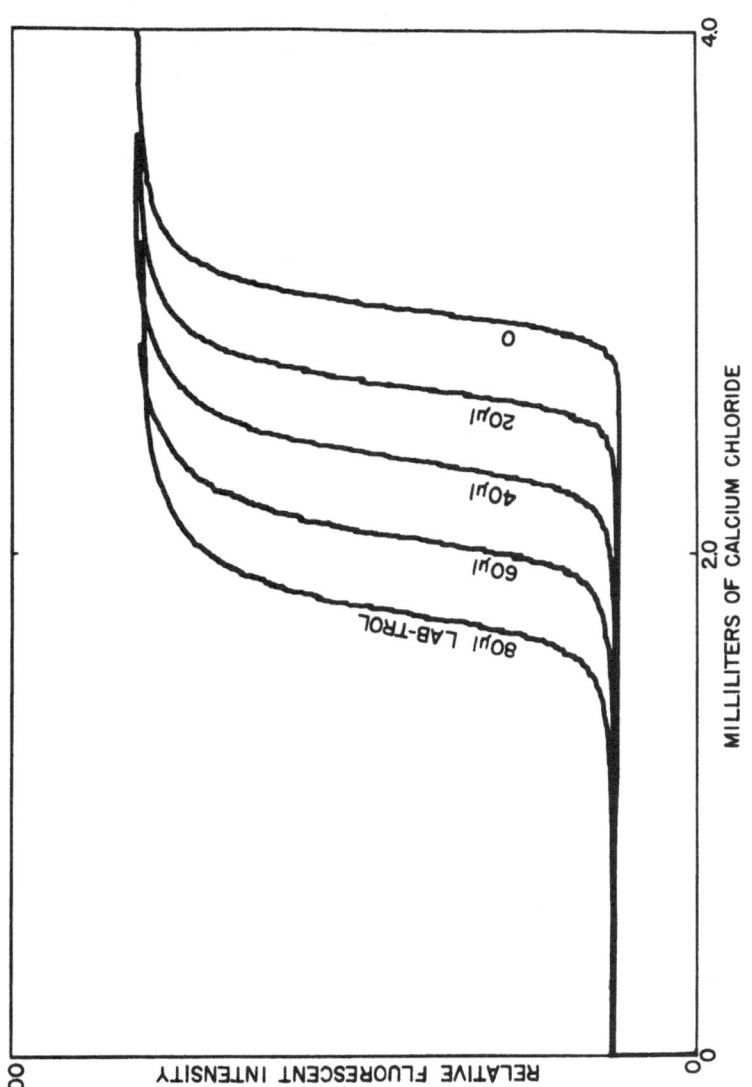

Fig. 8. Titration curves obtained when titrating 20 ml of 0.0004 M EDTA with 0, 20, 40, 60, and 80 ml of Lab-trol added to the titrate. The curves are parallel and equally spaced, therefore the concentration of calcium can be estimated from the spacing, provided the titrant concentration is known. The linear portions of the curves provide a convenient guide in determining the control-pointer setting when the automatic–stop method is used to obtain more accurate results.

Dynamic Titration Curves

Figure 8 shows curves obtained using 20 ml of 0.00004 M EDTA with 0, 20, 40, 60, and 80 μl of Lab-trol added to the titrate. Since all five curves are parallel and equally spaced, the automatic-stop method is suitable for determining the equivalence point. When using the automatic-stop method, it was found that the most accurate results were obtained by setting the control pointer so that the titration was stopped in a region contained within the linear portion of the titration curve.

Automatic-Stop Method

Titrant: Harleco* calcium chloride concentrate, Item No. 3828, was diluted with H_2O to a final concentration such that 1 ml was approximately equal to 1 mg of $CaCO_3$. (Calcium concentration of 400 μg/ml.)

Titrate: Harleco EDTA, Item No. 3854, was diluted with H_2O to a concentration such that 1 ml was approximately equal to 0.02 mg of $CaCO_3$. (Calcium equivalent of 8 μg/ml). To 800 ml of this solution was added 40 ml of 1%, by weight, NaCN (aqueous), 40 ml of 5 N NaOH, and 10 ml of Calcein solution (made by dissolving 0.25 g of Calcein in 4 ml of 1 N NaOH and 30 ml H_2O and diluting 1 : 100 with H_2O). One milliliter of this solution contained a calcium equivalent of 7.2 μg. Twenty milliliters was used for each titration.

Instrumental Parameters

Meniscomatic buret: 3.5 ml capacity.

Titration vessel: 30-ml beaker.

*Hartman-Leddon Co., Philadelphia, Pennsylvania.

Difference current: 2 μa (4 μa initial, with automatic stop at 6 μa).

Results

Table II shows the results of titrating various quantities of titrant solution, Lab-trol, and Patho-trol.

Table II. Determination of Calcium in Lab-trol and Patho-trol.

Trial		Calcium added, μg			Calcium found, μg	Error, %
		0	40*	100†		
Titrant	1	0.362	0.262	–	40	0
	2	0.366	0.262	–	41.6	+4
	3	0.363	–	0.115	99.2	–1
	4	0.364	–	0.113	100.4	0
	5	0.364	–	0.120	97.6	–2
		0	50‡	100§		
Lab-trol	1	0.362	0.234	–	51.2	+3
	2	0.365	0.231	–	53.6	+7
	3	0.363	0.234	–	51.6	+3
	4	0.362	–	0.112	100	0
	5	0.366	–	0.113	101	+1
	6	0.362	–	0.112	100	0
	7	0.362	–	0.116	98.4	–2
		0	40.5"	81.0#		
Patho-trol	1	0.368	0.256	–	44.8	+11
	2	0.365	0.265	–	40.0	–1
	3	0.368	–	0.164	81.6	+1
	4	0.365	–	0.163	80.4	–1

Twenty milliliter of EDTA solution (1 ml equivalent to 72 μg calcium) were titrated using Calcein as the indicator of fluorescence end point. The titrant contained 400 μg calcium/ml and the entries in the table represent milliliters of this titrant added for automatic stop. Dade Reagents analysis was 100 ± 4 μg/ml for Lab-trol, and 81.0 ± 3.2 μg/ml for Patho-trol. Procedure and instrumental parameters are given in the text.

*0.1 ml titrant   ‡0.5 ml Lab-trol
†0.5 ml titrant diluted 1:2 with H_2O §1.0 ml Lab-trol
                                          "0.5 ml Patho-trol
                                          #1.0 ml Patho-trol

The former furnished an obvious check on accuracy, since no accuracy greater than that achieved by titrating the titrant could be expected. The determinations listed in Table II are representative. The results agree favorably with those obtained by Toribara and Koral [7] (using visual detection of the fluorescence end point and oxalate precipitation of calcium), and by Ashby and Roberts [8] (who used Lab-trol without oxalate precipitation of calcium and visual detection of fluorescence end point).

Calcium in Limestone

Sample

Three 1-g samples of argillaceous limestone* were dissolved in accordance with the procedure of Jordan and Billingham [9]. The N.B.S. analysis was: Al_2O_3, 4.16%; Fe_2O_3, 1.63%; MgO, 2.19%, and CaO, 41.32%. From this analysis the final calcium concentration in the sample solution was 590 μg/ml, computed from the dissolution procedure in which 500 ml of solution contained the 1-g limestone samples.

Titrant

Harleco calcium chloride concentrate was diluted as previously described to a calcium concentration of 400 μg/ml.

Titrate

Harleco EDTA concentrate was diluted as previously described, except that 1 N instead of 5 N NaOH was used. One milliliter of this solution contained a calcium equivalent of 18 μg. Twenty milliliters was used for each titration.

---

*Standard sample No. 1a, National Bureau of Standards, Washington, D.C.

Instrumental Parameters
   Same as given previously.

Results
   Table III shows the results of titrating 0.5 ml of each
   limestone solution. The mean relative error of −6%
   indicates about a −4% error in the dissolution procedure, since the instrumental error amounted to
   approximately 2%. The determinations listed are
   representative. Variation in instrumental parameters, such as titrant delivery rate (5 : 1), difference

Table III. Determination of Calcium in N.B.S.
Limestone 1a

Trial	No calcium added	Titrant added, 0.5 ml (200 μg Ca)	Limestone solution added, 0.5 ml		
			Sample A	Sample B	Sample C
1	0.9392	0.4163	0.2457	0.2000	0.2128
2	0.9111	0.4086	0.2318	0.2234	0.2112
3	0.9089	0.4089	0.2487	0.2135	0.2272
4	0.9154	0.3898	0.2344	0.2210	0.2370
5	0.9118	0.4104	0.2514	0.2217	0.2249
Arithmetical mean	0.917	0.407	0.242	0.216	0.223
Ca found, μg	–	204	270	281	278
Ca from analysis, μg	–	200*	295†	295†	295†
Error,‡ %	–	+2.0	−8.5	−5.0	−6.0

Twenty milliliters of EDTA solution (1 ml equivalent to 18 μg calcium) was
titrated with Calcein as the indicator of the fluorescence end point. The
titrant contained 400 μg calcium/ml and the entries represent milliliters of
this titrant added for automatic stop. The procedure and instrumental parameters are given in the text.

*From titrant concentration.
†From N.B.S. analysis and dissolution procedure.
‡The consistently low values obtained are probably due to loss of calcium
in the dissolution procedure.

current (4 : 1), and stirring rate, did not improve the accuracy. The parameters were easily varied, since about 40 titrations were performed in 1 hr.

ACKNOWLEDGMENT

We appreciate the assistance of Marjorie Wheeler in preparing the solutions and performing many of the titrations.

REFERENCES

1. Healy, S. U.S. Patent 2,925,198 (to American Instrument Co., Inc.), Feb. 19, 1960.
2. Klaasse, J. M. Paper presented at the 14th Annual Instrument-Automation Conference and Exhibit, Chicago, Illinois, Sept. 1959. Reprint No. 114, available from American Instrument Company, Inc., Silver Spring, Maryland.
3. Mosteller, F. Ann. Math. Statistics, Vol. 17, p. 377, 1946.
4. Diehl, H., and Ellingboe, J. L. Anal. Chem., Vol. 28, p. 882, 1956.
5. Howerton, H. K. I.S.A. J., Vol. 6, p. 50, 1959. [See also, U.S. Patent 2,971,429 (to American Instrument Company, Inc.) Feb. 14, 1961.]
6. Clark, E. P., and Collip, J. B. J. Biol. Chem., Vol. 63, p. 461, 1952.
7. Toribara, T. Y., and Koral, L. Talanta, Vol. 7, p. 248, 1961.
8. Ashby, R. O., and Roberts, M. J. Lab. and Clin. Med., Vol. 49, p. 958, 1957.
9. Jordan, J., and Billingham, E. J., Jr. Anal. Chem., Vol. 33, p. 120, 1961.

Titrations in Nonaqueous Solvents

NONAQUEOUS TITRATIONS

J. S. Fritz
Iowa State University, Ames, Iowa

The study of nonaqueous titrations covers a very broad field. It can include acid-base titrations, oxidation-reduction titrations, precipitation-titrations, and titrations involving complex formation. In order to limit things somewhat, I would like to discuss only nonaqueous acid-base titrations. However, I do not mean to imply that other types of titrations carried out in nonaqueous media are of lesser importance. For example, some excellent oxidation-reduction titrations have been carried out by Dr. Stone of Michigan State and as an example of a precipitation titration, we developed a method several years ago for titrating sulfate with barium that is carried out at least partly in nonaqueous solution. Other examples could be given but to mention all of them would lead to an excessively prolonged discussion.

There are two major areas of use for nonaqueous acid-base titrations. The first is the direct determination of compounds that have definite acidic or basic properties. This includes many thousands of organic compounds. For example, many amines are basic and can be titrated. There are also numerous acidic compounds, such as sulfonic acids, phosphonic acids, carboxylic acids, enols, imides, phenols, sulfur compounds, and others which can be titrated. The second largest and most important use of nonaqueous acid-base titrations is the indirect deter-

mination of organic functional groups. This is an extremely important field of analysis. For example, if you wish
to determine carbonyl compounds, the best way is to form
the oxime and then determine the excess hydroxylamine,
either as the oxime or the hydroxyl ammonium chloride.
This is best done in an almost completely nonaqueous
system.

What are some of the advantages of nonaqueous acid-
base titrations? First of all, they are cheap. Secondly,
they are simple; in many cases you don't need a trained
chemist to do them. A third advantage is that nonaqueous
acid-base titrations are extremely accurate. For the
analysis of organic compounds where you are assaying a
major constituent, nonaqueous acid-base titrations are
unmatched in their accuracy. I think it is well to reme-
ber that some of the very excellent instrumental methods
are, in many cases somewhat inaccurate. Finally, why use
some smelly solvent such as acetic acid or pyridine, as is
frequently necessary, instead of water or perhaps water
mixed with alcohol? The reason for this is that nonaqueous
titrations are much broader in their application than are
acid-base titrations carried out in water or in water mixtures with a waterlike solvent such as alcohol.

Now I would like to divide the talk roughly into two
parts: first, the determination of bases and second, the
titration of acids. The first topic under the determination
of bases is acid-base equilibria in glacial acetic acid.

If you want to titrate a weak base you need a solvent
that is not basic itself. Water is too basic to titrate a
very weak base and alcohol is also basic. A solvent that
does not have appreciable basic properties such as glacial acetic acid, acetonitrile, nitromethane, chloroben-
zene etc., should be used. If you wish to titrate a weak

base, the titrant should be as strong an acid as possible. In water it doesn't make too much difference which strong acid you choose, because water levels the common acids to approximately the same strength and you are really titrating with the H_3O^+ ion. In glacial acetic acid and other solvents it does make a difference. It turns out that perchloric acid is much better than other acids for the titration, so that in glacial acetic acid solvent, perchloric acid titrant is probably the most commonly used. Bruckenstein and Kolthoff have made extensive theoretical studies on equilibria in glacial acetic acid. I would like to briefly discuss and review some of their work, because it gives an insight into the type of equilibria encountered in nonaqueous solvents. It also gives some idea of just how the behavior of acids and bases in nonaqueous solvents differs from their behavior in water.

Glacial acetic acid has a rather low dielectric constant; Bruckenstein gives it a value of about 6. In a solvent with such a low dielectric constant, virtually all acid-base reactions involve extensive ion-pair formation. Reaction 1 shows a general equilibrium, where HX represents a

$$HX + HAc \rightleftharpoons H_2Ac^+ X^- \rightleftharpoons H_2Ac^+ + X^-$$

(Reaction 1)

rather strong acid which might be either hydrochloric acid or perchloric acid. First, the acid reacts with the solvent to form the ion pair—the reaction we call ionization. Then the ion pair dissociates to form the free ion—the reaction we call dissociation. Two constants characterize the over-all reaction, the ionization constant and the dissociation constant. It is convenient, instead of working with two constants, to work with an over-all constant. To simplify matters further, it is convenient to talk about the free proton, H^+, instead of the solvated proton, H_2Ac^+. Thus, the over-all dissociation constant for strong acids is given by the relationship

$$K_{HX} = \frac{[H^+] \, [X^-]}{C_{HX}}$$

where $[H^+]$ is the concentration of free hydrogen ions (really solvated), and $[X^-]$ is the concentration of the free anion, and C_{HX} is the sum of the concentration of undissociated acid and the concentration of ion pairs.

When one considers the ionization of a salt, such as an amine perchlorate dissolved in glacial acetic acid, a third type of constant is required. The ionization of the salt in acetic acid must here be considered. This is different from the case where a salt is dissolved in water, because one ordinarily considers salts to be completely dissociated in water.

Having defined some of these over-all constants, it might be useful to examine their magnitude. In Table I, values are listed for a number of acids, bases, and salts. Notice that the acids are far from completely dissociated. The strongest acid, perchloric acid, seems to have about the same strength in acetic acid as acetic acid does in water. The other acids are even weaker. The bases stack up somewhat differently. Pyridine, for example, has a more favorable constant in glacial acetic acid than in water. To make a rough generalization, bases are about as strong in glacial acetic as they would be in water. Notice that salts are far from completely ionized and exist mostly as ion pairs. These constants, incidentally, were determined mainly spectrophotometrically. A few were verified or determined potentiometrically, or by conductive measurements.

Table I. Over-all Dissociation Constants of Acids, Bases, and Salts in Glacial Acetic Acid (-log autoprotolysis constant of glacial acetic acid, $pK_S = 14.45$).

		pK_{HX}
Acids	Perchloric acid	4.87
	Sulfuric acid	7.24
	p-Toluenesulfonic acid	8.46
	Hydrochloric acid	8.55
		pK_B
Bases	Tribenzylamine	5.36
	Diethylaniline	5.78
	Pyridine	6.10
	Potassium acetate	6.10
	Sodium acetate	6.58
	Lithium acetate	6.79
	2,5-Dichloroaniline	9.48
	Urea	10.24
		pK_{BHX}
Salts	Sodium perchlorate	5.48
	Diethylaniline perchlorate	5.79
	Tribenzylamine hydrochloride	6.71
	Potassium chloride	6.88
	Urea hydrochloride	6.96
	Lithium chloride	7.08
	Dodecylamine hydrochloride	7.45

Reaction 2 summarizes the equilibria which occur when a base is titrated in glacial acetic acid with per-

(Reaction 2)

$$
\begin{array}{ccc}
HClO_4 & B & \\
\updownarrow K_{HClO_4} & \updownarrow K_B & \\
H^+ & + \quad A_c^- & \overset{1/K_s}{\rightleftharpoons} \quad HA_c \\
+ & + & 1/K_{BHClO_4} \\
ClO_4^- & + \quad BH^+ & \rightleftharpoons \quad BH^+ClO_4^-
\end{array}
$$

(Over-all Reaction) $HClO_4 + B \rightleftharpoons BH^+ClO_4^- + HA_c$
(ion pair)

(Equilibrium Constant) $K = \dfrac{C_{BHClO_4}}{C_{HClO_4} C_B} = \dfrac{K_{HClO_4} K_B}{K_{BHClO_4} K_S}$

chloric acid as the titrant. The base 'B', when dissolved in the acetic acid, equilibrates to form some acetic ion and some BH^+ ion. The equilibrium constant for this is K_B, the over-all constant previously described. Perchloric acid is considered as undissociated. Its dissociation is an ionization, and is represented by the equilibrium constant K_{HX}. The hydrogen and acetate ions combine. The constant for this is $1/K_S$, K_S being the autoprotolysis constant for glacial acetic acid. K_S is equivalent to K_W in water, and furthermore it happens to have just about the same value as K_W. K_W is 10^{-14} and K_S is $10^{-14.5}$, but since equilibrium is in the opposite direction, K is $1/10^{-14.5}$, or $10^{+14.5}$. In addition, BH^+ and ClO_4^- combine to form an ion pair in the salt.

How does the situation in glacial acetic acid compare to that in water You start off at a disadvantage because the dissociation constant of the acid is only of the order of 10^{-5}. The dissociation constant of the base is roughly the same in acetic acid as it is in water. We have seen that the autoprotolysis constant in glacial acetic acid is about the same as in water. You gain, however, because K_{BHX}, which represents the equilibrium involving ion-pair formation, is about 10^5 for this reaction. So you end up about as well off in glacial acetic acid as you do in water, in spite of the fact that you are titrating with an acid that is only weakly dissociated. Reaction 2 summarizes the overall reaction.

Suppose we consider one or two examples to see what happens. The equilibrium constant for the titration of pyridine in glacial acetic acid is 10^{-6}, which means that we end up with a formation constant of $10^{9.5}$. How does this compare with water? In water the formation constant is $10^{5.1}$. Therefore, it is much better to titrate pyridine in glacial acetic acid than in water. The contrast is even greater with urea, which is a very weak base. The titration of urea is very marginal, even in glacial acetic acid, and is impossible to titrate urea in acetic acid; therefore, calculating the formation constant, we arrive at $10^{5.4}$. This is a very marginal value if one is to achieve a good end point.

If the data are available, one can predict the feasibility of any titration quite accurately. Unfortunately, the necessary data are not available for a great many bases. However, Streuli and others have shown that there is a relationship between the strength of a base in glacial acetic acid or any other nonaqueous solvent, and the strength of the same base in water. This means that if one has the necessary constants for just a few bases in both types of solvents and can predict how they will titrate in glacial acetic acid, then, knowing the constants of other bases in water, one can extrapolate their behavior to acetic acid, even though the glacial-acetic-acid constants are not available.

Suppose we consider a few practical features concerning the titration of bases. How are they titrated? A simple visual titration will do. Just add an indicator and titrate until you get an end point. This is just as simple and accurate as similar titrations in water. You can titrate potentiometrically, and it is a good idea to try a potentiometeric titration before you use an indicator. Again, this is just as simple as in water. You use a pH meter with glass-calomel electrodes. You don't use the pH scale, you use the millivolt scale. Amines of all types can be titrated. Aliphatic amines are very strong bases and are very easy to titrate. Aromatic amines can be titrated, unless they are heavily substituted with electron-withdrawing substituents, such as chloro, bromo, or nitro groups. Nitrogen heterocyclics such as pyridine, and quinoline are sufficiently basic to be titrated. Some polymers have basic groups and, providing that they can be dissolved, there is an excellent chance that they can be titrated. Amino acids can be titrated, although they are sometimes a little difficult to dissolve. Amine salts are of considerable interest to the pharmaceutical industry. An amine nitrate can be titrated directly in glacial acetic acid:

$$RNH_3^+ NO_3^- + HClO_4 \rightarrow RNH_3^+ ClO_4^- + HNO_3$$

You get nitric acid as the neutralization product. Nitric acid doesn't sound like much of a neutralization product. but in glacial acetic acid it is quite a weak acid. You get a nice neutralization curve (potentiometrically) with a break of about 200 mv. You can also titrate amine sulfates, providing you can get them into solution. You have to use some trickery when titrating amine hydrochlorides, since hydrochloric acid is not a weak acid in glacial acetic acid. What you do is add mercuric acetate—this gives undissociated mercuric chloride, which is soluble but undissociated—plus the amine acetate. Now the amine acetate can be titrated with perchloric acid in a straightforward manner. But, you may ask, what about mercuric acetate? Don't you have to worry about that? It turns out that this is also undissociated and does not titrate as a base, provided you use a modest excess.

Suppose we consider some very weakly basic compounds, such as ketimines, phosphines, and oxiranes. A very interesting method of dealing with oxiranes was developed by Durbetaki. The oxirane was reacted with hydrobromic acid to form the bromohydrin. This type of reaction has long been known using hydrochloric acid, but in that medium the reaction takes approximately three hours. In glacial acetic acid, the reaction is enough to allow you to titrate directly at normal speed. You can get an end point potentiometrically or with an indicator. In fact, if you have a mixture of amine and oxirane, you can get two potentiometric breaks, the first for the amine and the second for the oxirane. Amides, phosphene oxides, triphenyl methanol, and amine oxides are very weak bases and cannot be titrated in glacial acetic acid under ordinary conditions. However, they can be titrated if one uses acetic anhydride as solvent, or if one uses a solvent that is mixed with acetic anhydride. Why does acetic anhydride work? There are two reasons. First, it removes the last trace of water from the solution; secondly, perchloric acid in the presence of acetic anhydride forms the ion CH_3CO^+. Since this is an extremely reactive substance, one can titrate very weak bases.

In addition to the determination of total base, it is also possible to titrate mixtures. This can be done in two ways. One is to titrate mixtures based on the type of amines present. For example, one can distinguish between primary, secondary, and tertiary amines. This is done simply. Acetylate the primary and secondary amines in the mixture with acetic anhydride. They are converted to amides which are only weakly basic. Tertiary amines are not affected and titrate very well. A further differentiation can be made, however. The primary amine can be reacted with salicyl aldehyde to form a Schiff base. The secondary and tertiary amines are unaffected as far as basic strength is concerned, so that one can titrate the sum of secondary and tertiary amines. By these two titrations plus a determination of total amine, one can resolve the mixture. This approach works well for aliphatic amines, but not for aromatic amines.

Another method of analyzing a mixture of bases is to utilize the difference in the basicity of its components. As an example, let us take a mixture of aromatic and aliphatic amines. Since aliphatic amines are more strongly basic, one would expect to get a titration curve with two breaks, one for the aliphatic amine and one for the aromatic amine. However, you must not use glacial acetic acid for this titration because you will get a curve similar to curve 'B' in Fig. 1. In other words, you get one potentiometric end point for the sum of the two. The reason is that glacial acetic acid reacts with aliphatic amines to form the acetate ion, which has about the same basic strength as the aromatic amine. Glacial acetic acid levels these two amines to the same strength. What you have to do is employ a nonaqueous solvent like acetonitrile and titrate with perchloric acid dissolved in dioxane. If you do this,

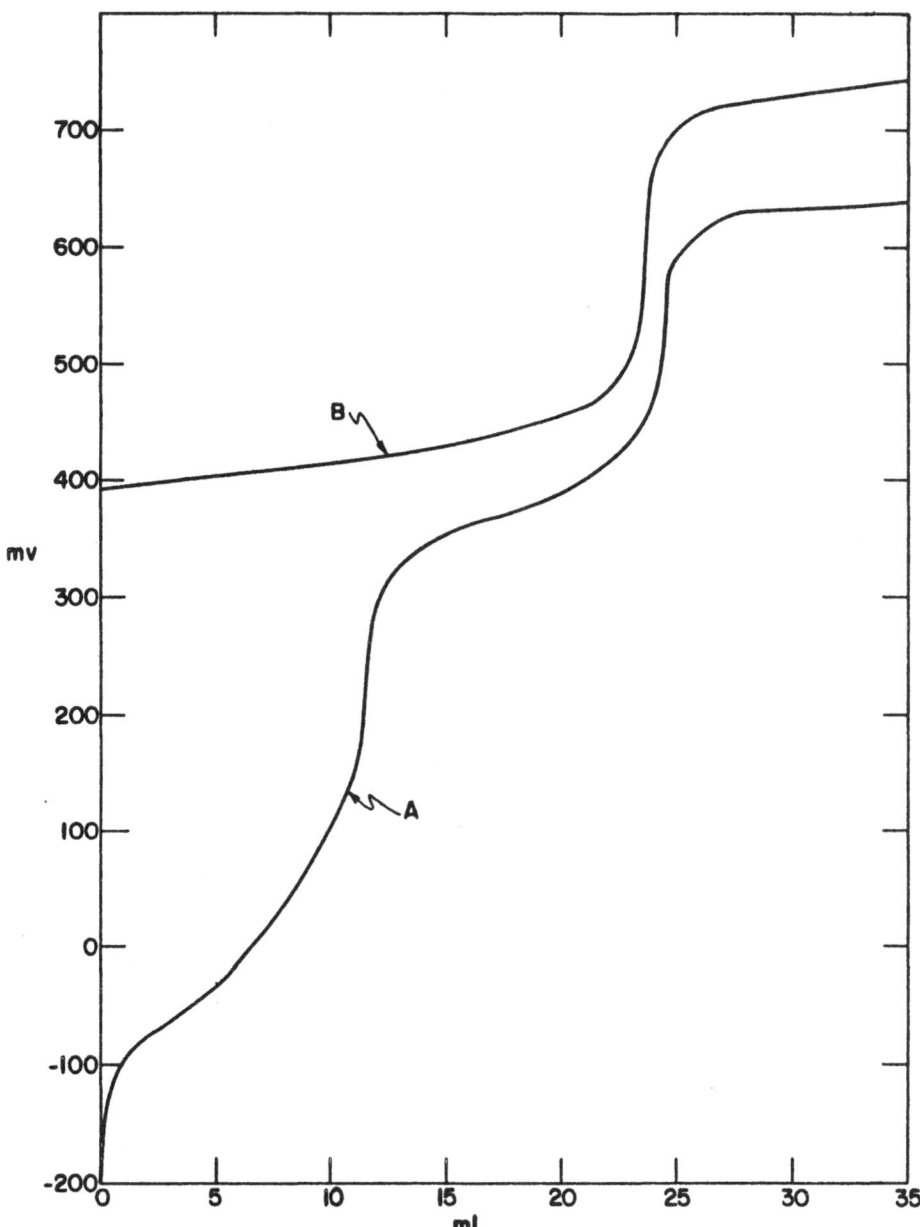

Fig. 1. Differentiating titration of tributylamine and N-ethylaniline in aceto-nitrile (Curve A). Curve B is the curve for titration of the same mixture in acetic acid.

you get a titration curve similar to Curve A in Fig. 1. Sometimes, when mixtures of weaker bases are being titrated, glacial acetic acid does not exert a leveling effect. This is true of a mixture of aniline and paranitroaniline, for which two breaks are observed on the titration curve.

One of the most interesting special methods for analyzing mixtures of weak bases was published recently by Hummelstedt and Hume. This example is concerned with a mixture of isomeric methyl nitroanilines. Figure 2 shows the result of a photometric titration of these compounds. The method, briefly, is to set the spectrophotometer for some arbitrary adsorption reading and choose a wavelength where one of the species adsorbs, but the other does not. The wavelength you choose is that of maximum adsorption by the weaker base. There are actually four species in solution, the two free bases and the protonated

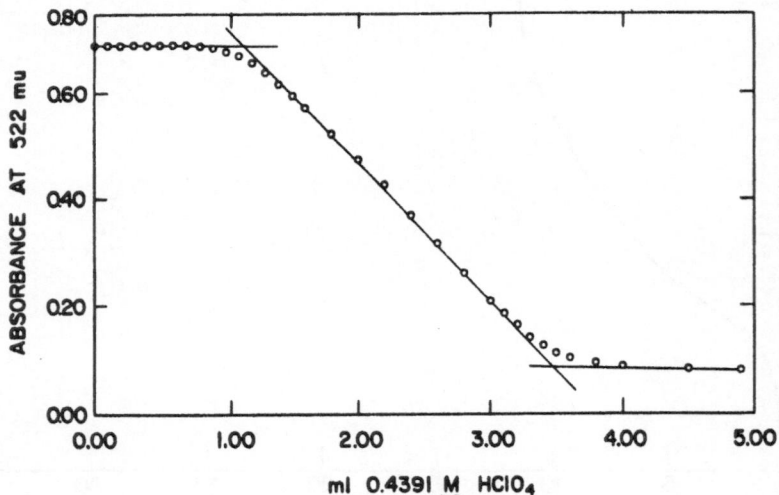

Fig. 2. Photometric titration of 2-methyl-5-nitroaniline and 4-methyl-2-nitroaniline. (Reprinted from Analytical Chemistry with permission.)

form of these bases. You choose the wavelength where only the free base of the weakest base adsorbs. When you begin to titrate, the stronger base will react preferentially and there will be no change in the adsorption. After the stronger base has been titrated the weaker base will react. This converts the free base to the protonated form and the adsorption will decrease in proportion to the neutralization. When neutralization is complete, the adsorption levels out so you obtain a nice differentiating titration with two breaks. Since the curve you obtain is mainly dependent on the accuracy of points well away from the end point, you will get better accuracy with this method than with a potentiometetric titration.

I would now like to consider the titration of acidic compounds in nonaqueous solutions. If you wish to titrate an acid in nonaqueous solution, you should choose a solvent that is not acidic and a titrant that is as strong a base as possible. The paper that really aroused people's imagination and created a lot of interest was the one published by Moss, Elliot, and Hall in 1948, in which they introduced ethylenediamine as a solvent. This compound certainly doesn't have any acidic properties and these authors showed that you can titrate phenol, which is normally too weak to titrate as an acid. In recent years, however, the trend has been away from the use of strongly basic solvents because they have a leveling effect on many bases and they are somewhat unpleasant to handle. Solvents now in use are pyridine, which is an inert solvent and a very weak base, acetonitrile, and acetone. Acetone and certain other ketones are surprisingly good. Recently we have done some work with tertiary butyl alcohol, an excellent solvent for certain cases. Sodium or potassium hydroxide can be used as titrants, but these are not particularly good if you are titrating a weak acid. It is much better to use sodium or potassium alkoxides, especially the methoxides. These are generally dissolved in a benzene—methanol solution (5 or 10 parts benzene to 1 part methanol). Why do you dissolve these in benzene-methanol instead of methanol alone? The reason is that methanol is somewhat acidic and the more methanol present, the less sharp the end point. Some methanol is necessary to dissolve the titrant.

The preferred titrant used by many people in recent years has been tetrabutylammonium hydroxide, also in benzene—methanol. This titrant has two major advantages over sodium and potassium salts. The first is that the products of titration are always soluble. With sodium and potassium you frequently get very gelatinous precipitates that are very disturbing. The second advantage of this titrant is that you can get excellent potentiometric curves using the glass electrode. With sodium and potassium titrants, the titration curves frequently do not tell the whole story. For example, you might get a titration curve with a very small break and yet when you add an indicator, the end point is extremely sharp. With tetrabutylammonium hydroxide you do not have this difficulty.

Reaction 3 shows the preparation of tetrabutylammonium hydroxide. There are several ways of preparing it,

$$Bu_4N^+X^- + Ag_2O + MeOH \rightarrow Bu_4N^+OH^- + Bu_4N^+OMe^- + 2AgX$$
(Reaction 3)

but I think that most people prepare it by the silver oxide reaction. You take the quaternary halide, dissolve it in a small amount of methanol, add silver oxide, and shake, and you get a precipitate of silver halide plus unreacted silver oxide, which is filtered off. You actually get a mixture of hydroxide and the methoxide. I felt instinctively that this was true and was gratified to find that Cluett had determined, by a very clever scheme, that this was the case. What he did was to take the mixture of these titrants and add glacial acetic acid to it. The glacial acetic acid reacted with the methoxide yielding methanol plus water from the hydroxide. He determined the water by the Karl Fischer reaction. By determining from another measurement the quantity of total base present, he was able to determine the amount of methoxide by difference. It turns out that there is approximately a 1:1 ratio of these two species. Although people call it tetrabutylammonium hydroxide, it is really a mixture of two titrants.

Again you can detect end points potentiometrically, using the glass-calomel electrode system and an ordinary pH meter. The only modification necessary is to replace the aqueous KCl in the reference electrode with methanolic KCl. You can also use visual indicators. Figure 3 shows some approximate transition ranges of these indicators dissolved in acetonitrile. These were determined by following the transition of these indicators, using the glass-calomel electrode. What you have here is roughly equivalent to the pH transition ranges of indicators in water. Unfortunately, it is quite difficult to reproduce potentials in nonaqueous systems and a certain amount of judgment is required.

Lets look at some sample titration curves to see what sort of characteristics they have. Figure 4 shows titration curves for a variety of phenols. Some of them are strong acids and some of them are weak. The addition of nitro groups adds to their acidic character. Note that some of these curves are quite flat while others have a very steep

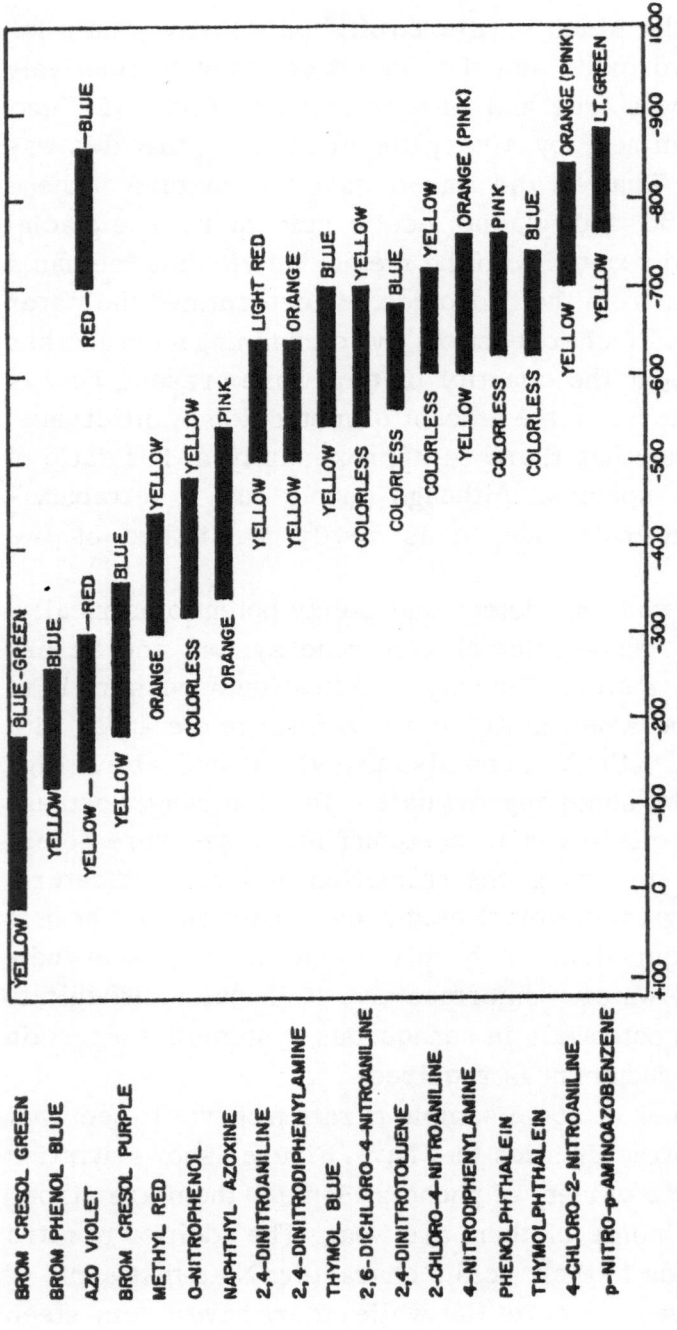

Fig. 3. Approximate transition ranges of indicators in acetonitrile.

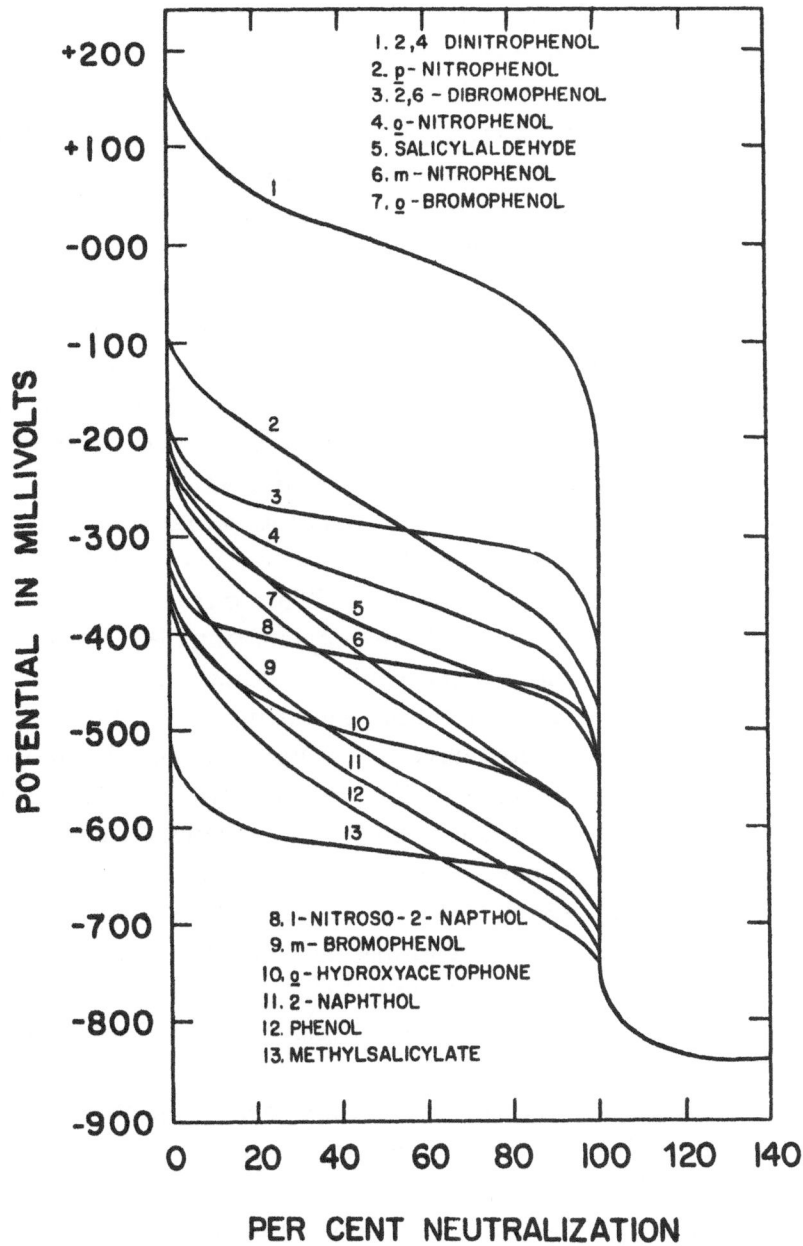

Fig. 4. Titration of phenols in acetone. (Reprinted from Analytical Chemistry with permission.)

slope. This was quite mystifying to us for a while, but was explained by Van der Heijde. His explanation is that you can have association between two molecules of phenol. Actually this association is between a molecule of free phenol and a phenolate anion (Fig. 5). This means that it is easy to titrate one hydrogen for every two molecules of phenol, after which this complex forms between phenolate anion and phenol. Thus, the curves with steep slopes indicate that you almost get an end point when you have added half a molecule of base for each molecule of phenol. This explains the curves with steep slopes, but what about the ones that are very flat, such as the curve for methyl salicylate? The explanation for the methyl salicylate curve is that you get the same sort of association complex, only it is within the molecule. How does this explain curve 3 (Fig. 4) for 2,6-dibromphenol? This curve is also quite flat. What is going on here? The explanation lies, I believe, in the large bromo groups sterically hindering association between two molecules and resulting in a flat curve.

Bruce and Harlow have presented additional evidence for this type of behavior. They did titrations in benzene, toluene, and gasoline. Figure 6 shows conductometric titrations in a solvent of very low dielectric constant. One of these phenols gives a conductometric curve with a very sharp peak at the point where half an equivalent of base has been added. The reason for the enhanced conductivity is that in solvents of low dielectric constant, these association complexes between the two molecules are greatly protonated. In the case of the other phenol, which is substituted in the 2,6- positions, this sort of association is sterically hindered.

The scope of acidic titrations in nonaqueous solvents is well illustrated by the behavior of succinic acid (Fig. 7).

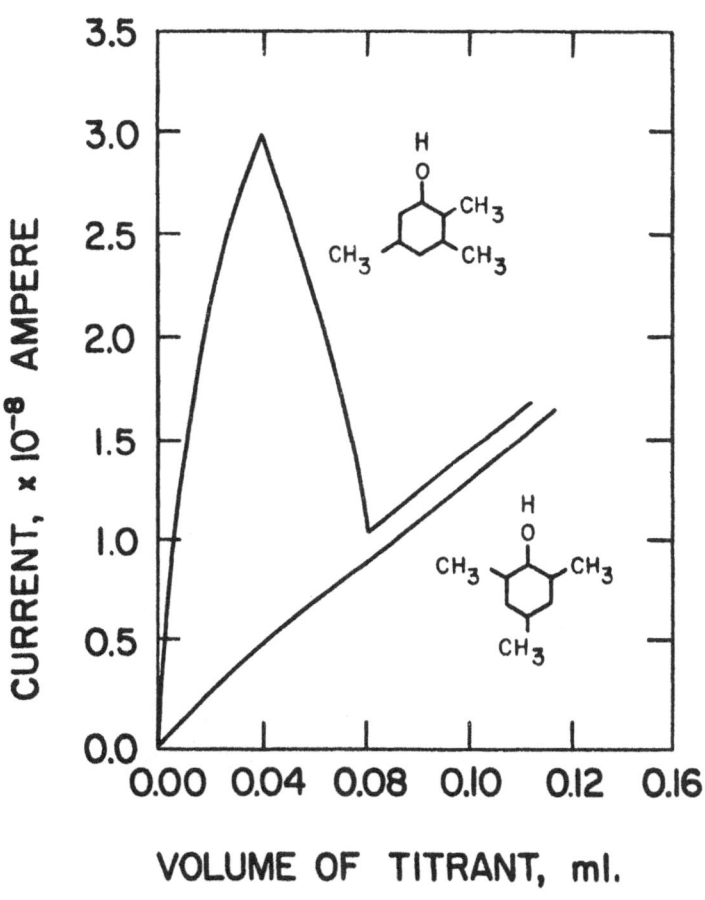

Fig. 5. Phenol association.

Fig. 6. Conductometric titration of phenols in toluene. (Reprinted from Analytical Chemistry with permission.)

Succinic acid is a dicarboxylic acid whose curve shows a break for each titrable hydrogen. What is interesting about that? Well, the acidity constants in water for succinic acid are very close to each other, so that you cannot resolve two breaks in a titration curve. You are

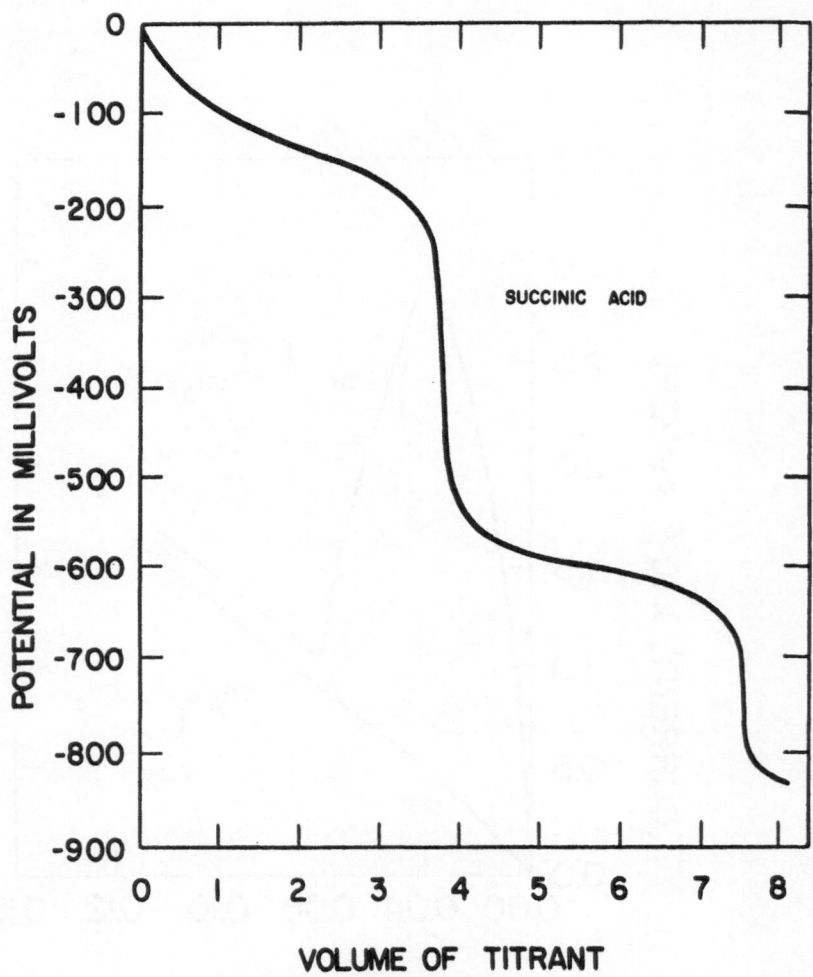

Fig. 7. Titration of succinic acid in acetone (after dissolution in a small amount of dimethylformamide). (Reprinted from Analytical Chemistry with permission.)

really living in a different world when you choose nonaqueous solutions. For example, you can actually titrate
sulfathrazole in the presence of sulfapyridine where the
difference in pK in water is about 1.3 units.

Strong acids can also be titrated easily. If you titrate
sulfuric acid, you get two tremendous breaks corresponding to the two available hydrogens. This is interesting
because it makes possible the analysis of mixtures of
sulfuric acid and sulfonic acid. If you titrate perchloric
acid in acetonitrile, you will get a potential break of about
1400 mv. Since each pH unit in water corresponds to about
59-mv change, this would correspond to a pH break of
24 pH units in an ordinary titration carried out in an
aqueous system. As I said before, you are living in a
different world when you do titrations in nonaqueous solutions.

COMPLEX FORMATION OF SILVER IONS WITH PHOSPHORUS AND NITROGEN ORGANIC COMPOUNDS IN ACETONE SOLUTIONS

C. A. Streuli
*American Cyanamid Company,
Stamford, Connecticut*

The exploitation of nonaqueous solvents for the analytical titration of organic bases and protonic acids has been intense in the last decade. Innumerable papers attest to the widespread interest in this field; the last review in "Analytical Chemistry" (April, 1960) listed some 223 entries for the 1958-60 period.

Besides the many publications on the practical aspects of nonaqueous work, a number of papers have demonstrated the theoretical basis of the method. In particular, the work of Bruckenstein and Kolthoff [1] has given a mathematical basis for the acetic acid system. Other workers [2-4] have demonstrated that a judicious choice of solvents can simplify a number of analytical problems. This choice appears to be dictated mainly on the basis of solute-solvent interaction. The problem of solvation is indeed one of the most interesting aspects of the entire field.

If the Brønsted view of acids and bases is abandoned in favor of the Lewis interpretation, a much wider application of titrants and nonaqueous solvents can be made to the acid-base field. In the Lewis view bases are still Brønsted bases—electron pair donors—but acids, instead of being confined to protons, include any electron pair acceptors. Thus all precipitation and complex-formation phenomena are included as well as the more conventional proton-base reaction. By the addition of the Lewis acids to the system, advantage may be taken for analytical purposes of the variation in reactivity of bases to various Lewis acids. Whether correlation exists between activity toward protons and toward other acids is a field worthy of further exploration.

The principal study in this field made in our laboratory during the past year has been on acid-base reactions between the Lewis acid, monovalent silver, and a number of bases, both ionic and molecular. All transition metals are, of course, Lewis acids and should undergo these reactions. The number of positions available for complex formation on a metallic ion will vary with the atomic number and can, in some instances, present a complex problem.

Monovalent silver was chosen for our studies for three reasons: its known ability to form nitrogen complexes in aqueous solutions; its well defined electrode reaction, suitable for potentiometric studies; and its low coordination number. Silver has an atomic number of 47 and a quantum structure:

$$1s^2;\ 2s^2;\ 2p^6;\ 3s^2;\ 3p^6;\ 4s^2;\ 3d^{10};\ 4p^6;\ 5s^1;\ 4d^{10}.$$

In the +1 state there are 3p and 1s orbitals available for coordination of donor groups. However, in most of the aqueous studies only 2 : 1 complexes have been noted, and X-ray studies show a linear complex utilizing an s and p orbital. If all three p orbitals are used for coordination as well as the s, a hybrid sp^3 tetrahedral structure is probably achieved.

EXPERIMENTAL

The perchlorate salt of silver is commercially available and shows good solubility in a number of nonaqueous solvents. Perchlorates themselves do not form stable complexes with metallic ions.

Acetone, the solvent of choice for these studies, is readily obtainable, cheap, and has wide solvent powers. A higher ketone might have been useful from the viewpoint of water content, but analysis of the purified solvent for water indicated less than 0.5% present. The acetone was purified by treatment with alkaline silver nitrate, followed by filtration and distillation. This procedure should remove aldehydes and other materials readily oxidized by silver.

Titrations were performed automatically, using either a Precision-Dow or Metrohm instrument and a silver billet combination electrode. Standard silver solutions were made from anhydrous silver perchlorate (G. Frederick Smith Chemical Company) dissolved in the purified acetone. They were approximately 0.1 N and were standardized against either standard sodium chloride titrated in aqueous solutions or diphenylguanidine in acetone. Samples of the bases which were titrated were usually 1- to 2-mM quantities dissolved in 100 ml of purified acetone.

RESULTS

Figure 1 shows the behavior of a number of bases in acetone when titrated with protons (0.1 N perchloric acid). Although each of the four species has at least one electron pair available to form a H^+:B bond, only the

nitrogen compounds are titratable. The substituted phosphine and iodide ions have such low proton affinities
that they cannot be titrated in acetone. They can, of
course, be titrated in other solvents (acetic anhydride).
It must be remembered in the case of both protons and
silver ion that the solvent is the limiting base (i.e., no
compound less basic than the solvent will titrate).

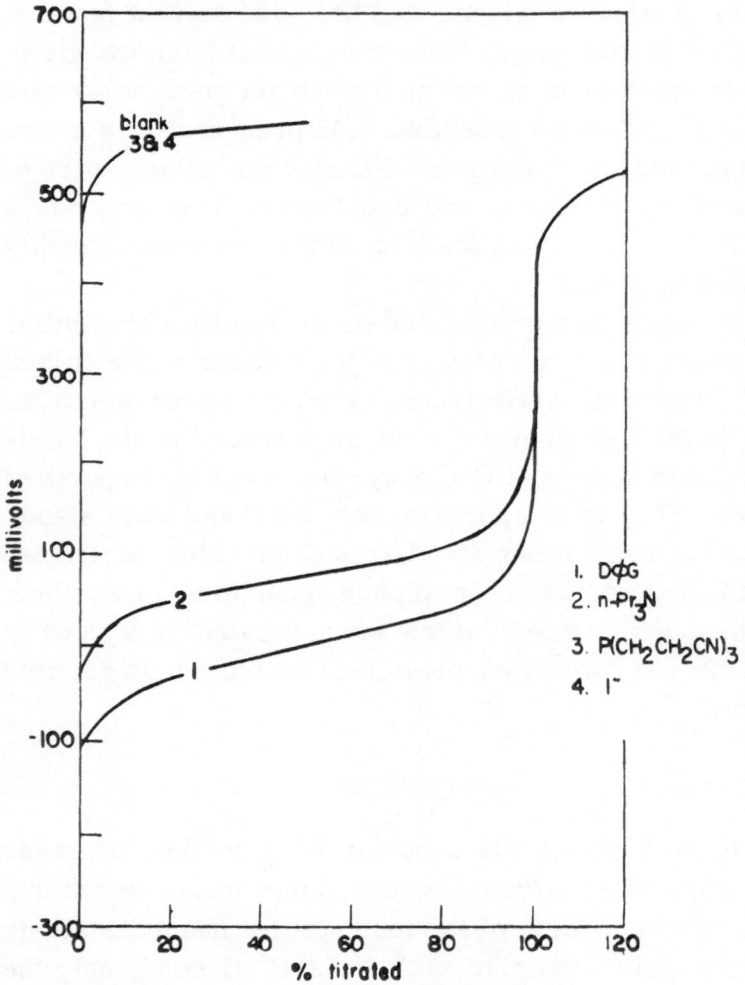

Fig. 1. Titration of bases in acetone with protons.

Figure 2 shows the behavior of these compounds when the Lewis acid, silver ion, is used. In this case all four bases are titratable and all show the formation of 2 : 1 complexes with silver ion. The order of basicity

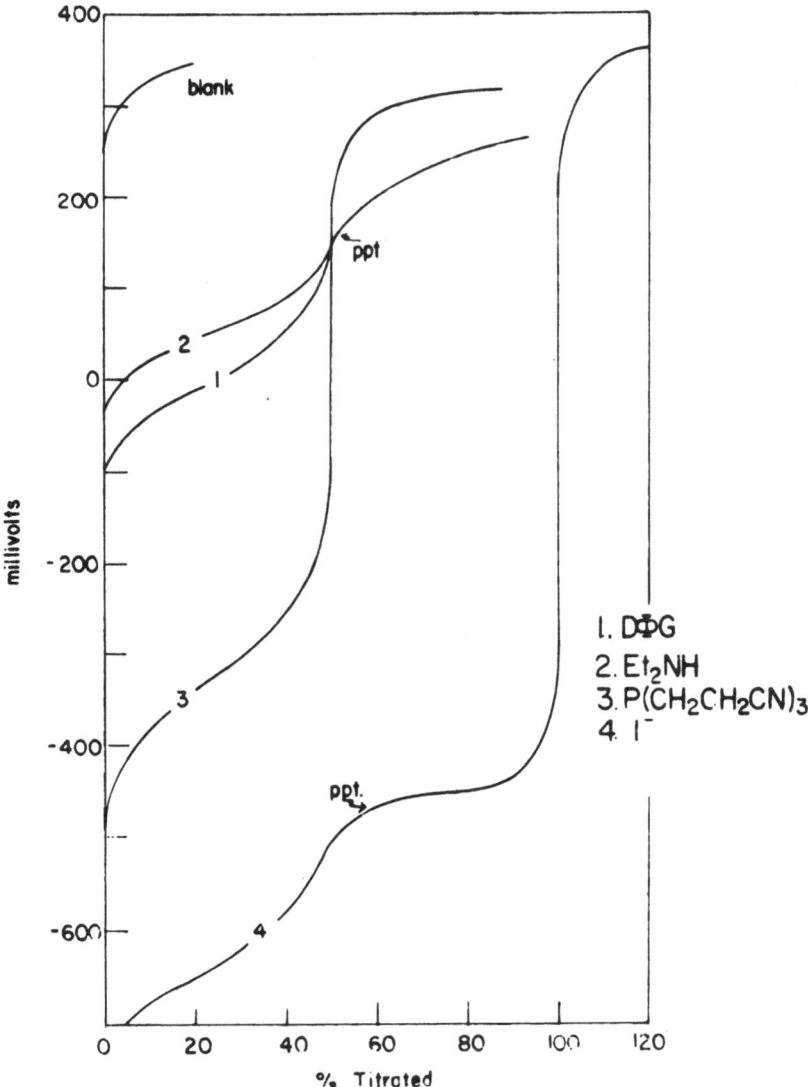

1. DΦG
2. Et$_2$NH
3. P(CH$_2$CH$_2$CN)$_3$
4. I$^-$

Fig. 2. Titration of Lewis bases in acetone with silver ion.

also changes. The guanidine and amine maintain their relative relationship, but the iodide ion shows the greatest basicity and the phosphine the next greatest. All the 2 : 1 complexes are soluble in acetone, and no precipitation of a compound occurs until the ratio of ligand to silver is less than 2 : 1. It is only after the half point of the titration that silver iodide and the secondary amine-silver compound precipitate. The guanidine and phosphine remain in solution throughout the titration.

The formation of stable soluble complexes has an analytical advantage over precipitation titration in that it avoids the problem of coprecipitation. In view of the rather unusual behavior of iodide ions in acetone, a short study of other inorganic anions which form insoluble silver salts in aqueous solutions was made for the acetone system. The results are shown in Table I.

Data on diphenylguanidine (DϕG) are included since this compound was used as a standard throughout the

Table I. Complex Formation in Acetone: Ag^+ and Inorganic Anions

Anion	2 : 1 complex	mcp	pK_a	pK_{sp}
DϕG	+	0	10.0	
I^-	+	− 632	−10.7	15.81
Br^-	+	− 472	− 7.7	12.11
Cl^-	−	− 325	− 4.7	9.81
CNS^-	^+i	− 285		12.31
CN^-	^+i	−1231	9.1	11.66

work. The silver-acetone system is in a sense unbuffered and the potential of the Ag^+-DϕG system, at a point mid-way between the initiation of the titration and the complete formation of the 2 : 1 complex, was assigned a potential of 0 mv in order to set up a reproducible relative scale. This potential is related to the pK of the complex species. By running DϕG every day, a check on the relative stability of the system could be maintained.

Of the five anions checked, only iodide and bromide form soluble complexes; the thiocyanate and cyanide complexes are insoluble, but are indicated by potential changes; chlorides precipitate immediately at the beginning of the titration. A consideration of the mid-complex potentials (mcp) indicates that in acetone the cyanide 2 : 1 complex is the most stable and the thiocyanate the least stable. The halides fall between these extremes; mcp values for the insoluble compound are not, of course, equilibrium values.

There are no indications of species other than 2 : 1 and 1 : 1 complexes. There is also no simple or obvious relation between mcp and pK_a of the bases. An inner relation for the halides does, however, appear to exist.

Since the AgI_2^- complex appears to be stable in relation to the precipitation of AgI in acetone while no such behavior is noticeable in water, a study of this behavior in mixtures of the two solvents was made and is shown in Fig. 3. As the water content of the solvent increases, precipitation occurs earlier and in 50% water is immediate. The lowered stability of the complex is shown in the disappearance of the inflection at the 2 : 1 iodide-silver ratio. The decrease in potential range is also reflected in the lowering of the excess-silver potential curve as the water content of the solvent increases. This is probably related to lower activities of Ag^+ ion in acetone.

The main body of this work was concerned with the complexes of silver ion and neutral molecular bases.

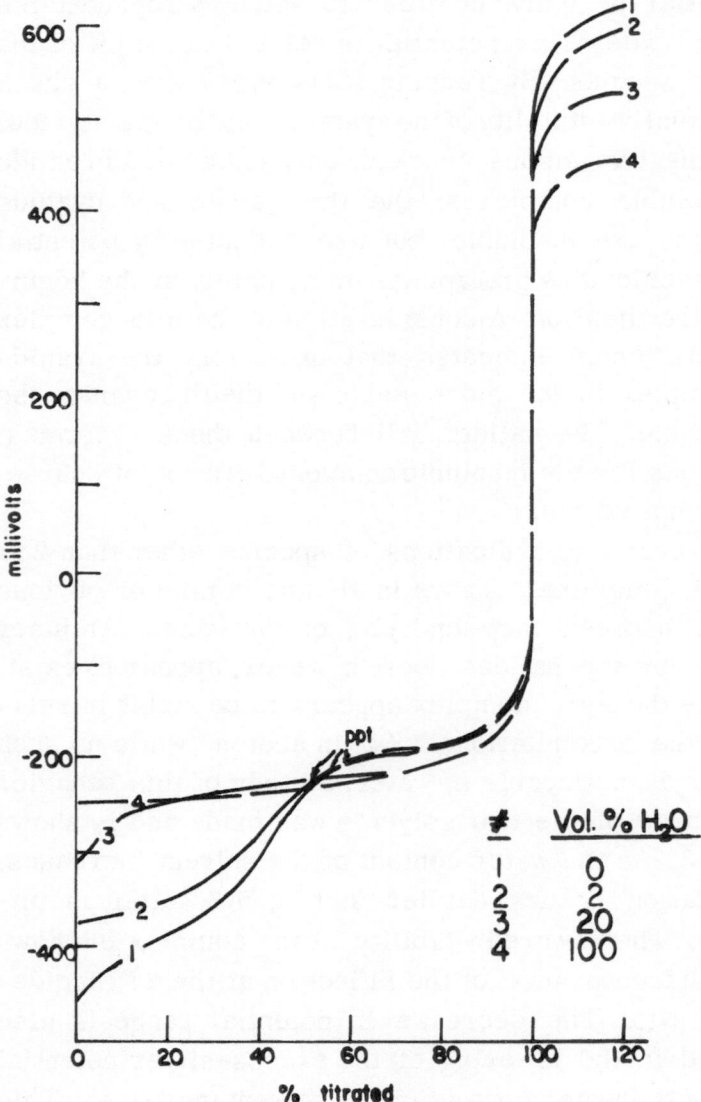

Fig. 3. Titration of potassium iodide with silver ion in acetone–water mixtures.

In this area nitrogen compounds are a logical first choice; Fig. 4 illustrates the titration behavior of a number of nitrogen compounds.

Curve 1 illustrates again the well-defined $2:1 D\phi G-Ag^+$ complex. No $1:1$ complex was detectable potentio-metrically. In all these cases 1 mM-meq to H^+ was the sample size taken. Primary amines such as n-propyl-amine give qualitative behavior similar to that of $D\phi G$:

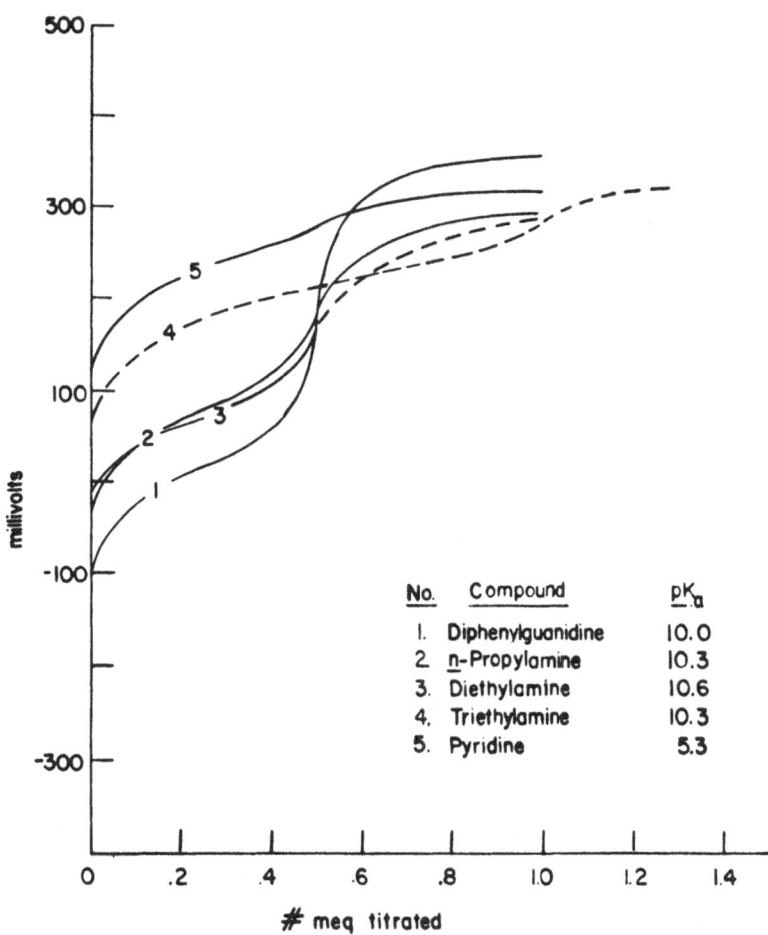

No.	Compound	pK_a
1.	Diphenylguanidine	10.0
2.	n-Propylamine	10.3
3.	Diethylamine	10.6
4.	Triethylamine	10.3
5.	Pyridine	5.3

Fig. 4. Complex formation of amines with silver ion in acetone.

a soluble 2:1 complex and no evidence of the 1:1 compound. The 2° amine, diethyl amine, also shows a 2:1
complex, but precipitation as represented by the dotted
line occurs after the rise in potential. The precipitate
was not identified. The tertiary amine solution precipitates at the beginning of the titration and does not indicate any 2:1 complex formation. The curve does,
however, indicate a 1:1 reaction with a weak inflection
of the titration curve. It may be noted that the difference

Table II. Stability of Silver–Amine Complexes in Acetone

| Compound | mcp (AgB_x) | | pK_a (H_2O) |
	1:2	1:1	
DΦG	0	288	10.0
n-$BuNH_2$	50	284	9.5
i-$BuNH_2$	68	316	9.4
2°-$BuNH_2$	−34	298	10.4
3°-$BuNH_2$	−82	298	10.7
n-Bu_2NH	21	(300)	10.2
i-Bu_2NH	119	(308)	9.8
2°-Bu_2NH	70	(327)	10.0
n-Bu_3N	(214)*	(236)	9.5
n-$PrNH_2$	63	255	10.3
Et_2NH	54	(240)	10.6
Et_3N	(159)	(193)	10.3
Pyridine	213	288	5.3

*Parentheses indicate nonequilibrium values.

in behavior for primary, secondary, and tertiary amines indicates the possibility of differentiating the first two classes from the third through the silver titration, even though proton affinities (pK_a values) are similar. The heterocycle, pyridine, may also be titrated and yields a rather weak break in the 2 : 1 region. Aromatic amines could not be titrated under the experimental conditions used.

The inability of the amines to form complexes with values greater than 2 : 1 probably reflects the relatively small size of the nitrogen atom, and strong polar forces. The absence of even a detectable 2 : 1 complex for the tertiary compound is most probably related to steric problems. A rather rough relation between mcp for 2 : 1 complexes and pK_a values also appears to exist from the relative positions of the curves.

Data of this type are recorded in Table II for a homogeneous group, the butyl amines. The qualitative behavior noted in the previous figure was preserved: primary amines formed soluble 2 : 1 complexes and did not precipitate even past the 1 : 1 point; 2° amines precipitated after the amine-silver ratio decreased below 2 : 1; the tertiary amine precipitated immediately. Mcp values for 2 : 1 and 1 : 1 (theoretical) complexes and pK_a values are listed. Bracketed values are nonequilibrium values. No correlation exists for the 1 : 1 mcp value and pK_a, but there is a relation for the 2 : 1 mcp values and the proton affinity. This is shown graphically in Fig. 5.

Separate relations exist for the primary and secondary amines, as shown by the two straight lines. The value for the tertiary amine appears to fall on the line for the secondary amines, but this may be fortuitous. The two relations may be expressed as simple linear equations and should be useful in predicting constants for the silver complexes from $pK_a(H^+)$ values. The steepness of the lines shows the reason why complexes of the more weakly basic aromatic amines are not observable in this solvent system. The two linear arrangements of the amines dif-

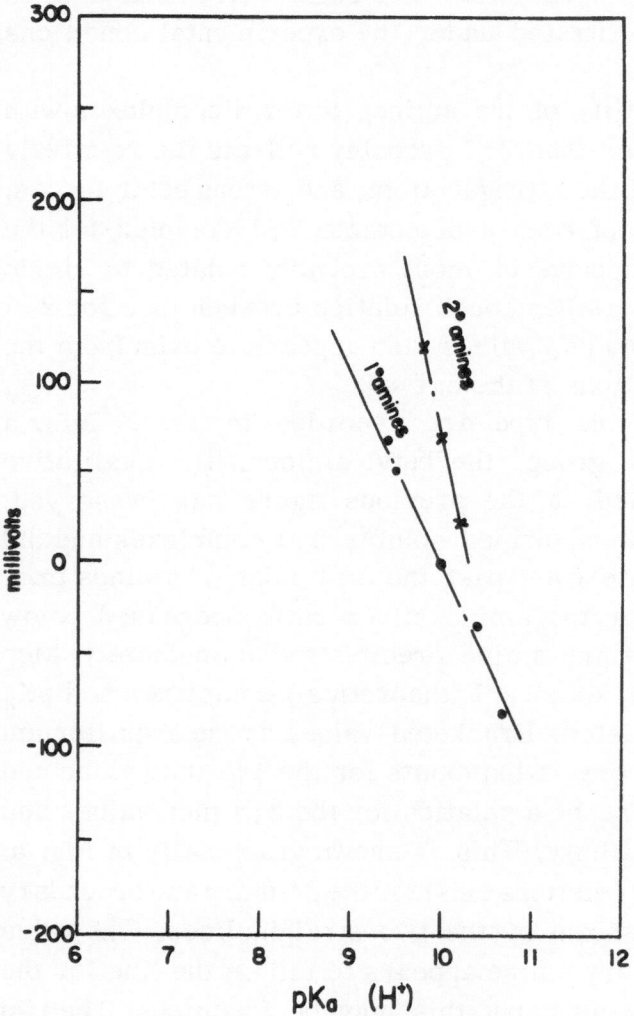

Fig. 5. Stability of silver—butylamine complexes in acetone vs pK_a (H+).

fering on the basis of type are not surprising in view of
aqueous work in this field and illustrate the strong de-
pendence of such relationships on structure.

A second group of nitrogen compounds of interest in
complex formation are the diamines. Titration curves
for several of these are shown in Fig. 6.

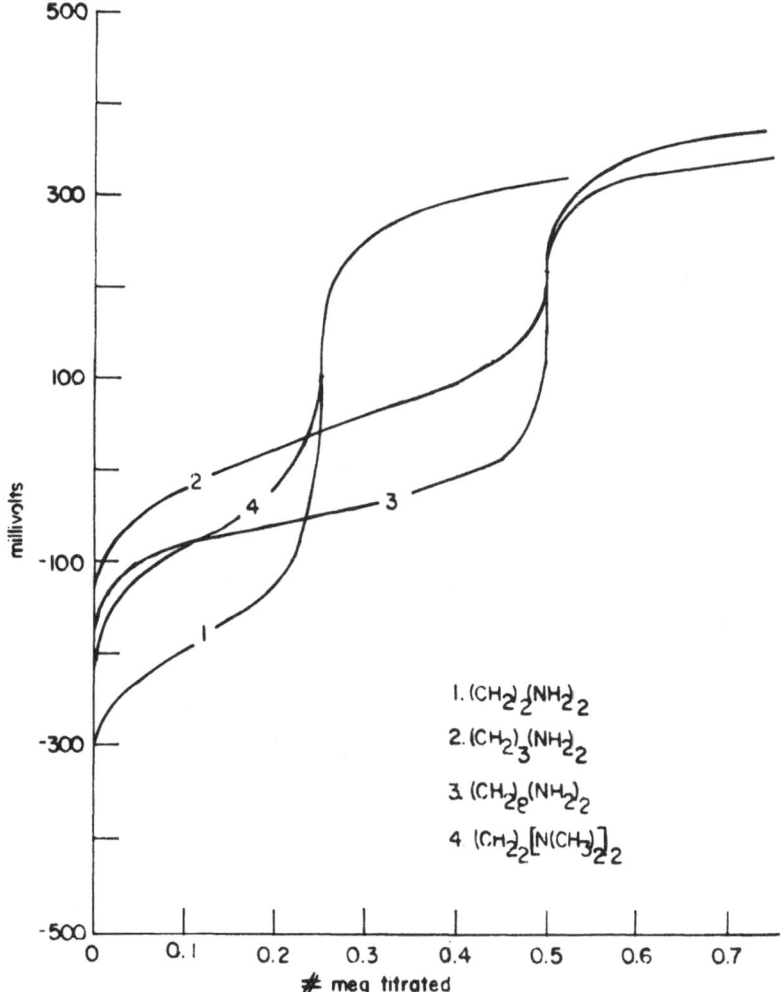

Fig. 6. Titration of diamines in acetone with silver perchlorate.

Two different types of complexes are noted, depending upon the length of the carbon chain between the two amine groups. Ethylenediamine (1) readily forms a stable 4 : 1 complex, the increased stability being undoubtedly due to chelation. Surprisingly, the tertiary derivative of ethylenediamine (4) also forms a stable 4 : 1 complex; chelation ability appears to have overcome the weak coordinating power noted for mono-functional tertiary amines. The stability of this latter complex is only a little less than that for the unsubstituted compound.

If the methylene chain is increased by one unit (2), the 4 : 1 complex does not appear to be stable. Only the 2 : 1 complex is shown in the titration behavior. However, the exact structure of the complexes are unknown. The octamethylene compound forms only 2 : 1 complexes and, presumably, so do the other diamines with more than three carbons in the chain.

Of all the nitrogen compounds studied in this work, only ethylenediamine and its derivative show the 4 : 1 complex which is theoretically possible. It should also be noted from the titration curves that the 2 : 1 octamethylenediamine complex (3) is more stable than the trimethylenediamine complex (2).

Another group of compounds, the substituted phosphines—phosphorus analogs of the amines—form an interesting class of compounds which have been studied by this technique. Phosphines are, in general, weaker proton-attracting bases than the corresponding amines, but show a gradation in base strength directly related to number and kind of substituents. The proton affinities of tertiary phosphines are greater than the secondary derivatives, which in turn are stronger than the primaries. This has been related to the larger phosphorus atom and lack of β strain in the molecules. The availability of d orbitals on phosphorus atoms for back-bonding with transition metals should lead to more stable silver— phosphine complexes than the corresponding silver— amine complex.

The titration behavior of a group of related substituted phosphines is shown in Fig. 7. The tertiary compound forms a well-defined 2:1 complex in contrast to the

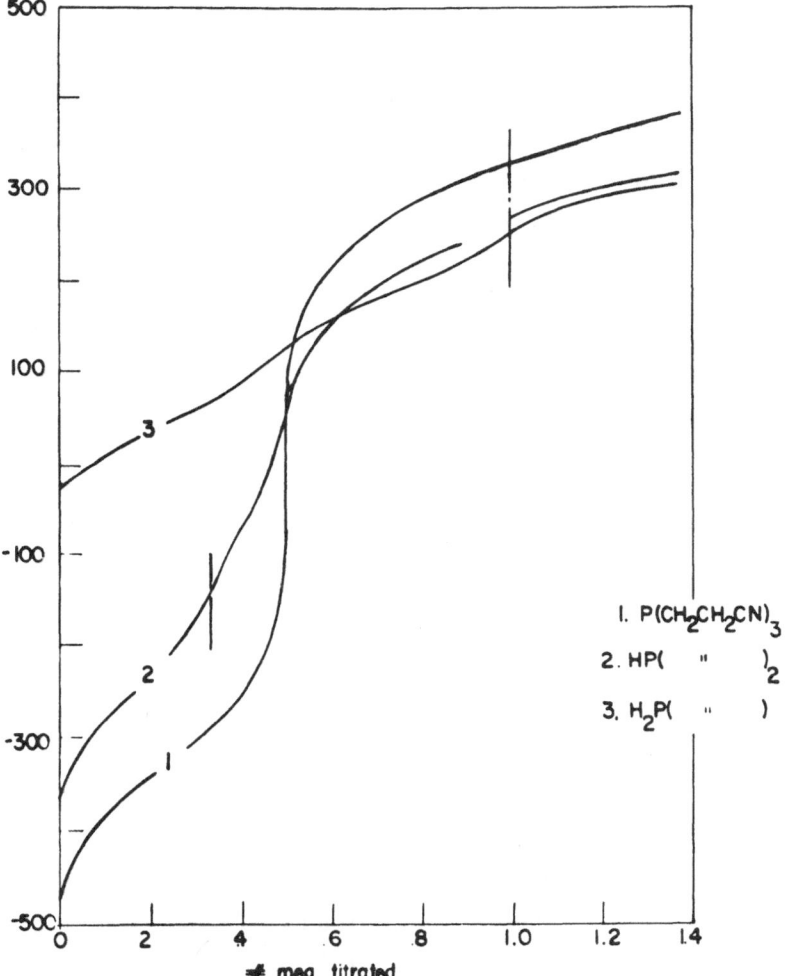

Fig. 7. Titration of substituted phosphines in acetone with silver ion.

tertiary amines and a weak 1 : 1 complex. The 2 : 1 complex is very stable with a mcp of -312 mv. This value for the 4 : 1 ethylenediamine derivative, the most stable nitrogen complex studied, is only -180 mv. The secondary derivative shows a 3 : 1 complex as well as the 2 : 1 and 1 : 1. Both of the higher complexes appear considerably more stable from potential data than does the 1 : 1. The 3 : 1 is presumably tetrahedral with either an acetone molecule in the fourth position or an empty orbital on the silver ion. The primary compound forms the least stable complexes. A 4 : 1 and 3 : 1 may exist, but are not well defined.

The order of complex formation is the reverse of that for amines but is consistent with $pK_a(H^+)$ data. All solutions remain clear throughout the titration, although a slight discoloration in the solution of the primary phosphine occurs with time. The back-bonding of the phosphorus atoms appear to be substantiated by the greater stability of the complexes.

Phosphines are, in general, oxidized by atmospheric oxygen and this process is evidently catalyzed in the presence of silver ion. However, if the acetone solutions are 0.1–0.2% in hydroquinone, quantitative titrations can be obtained. Hydroquinone itself is not titrated in this system.

The inability of the tri(2-cyanoethyl)phosphine to form 3 : 1 complexes with silver may be due to its bulk. Therefore, several other tertiary phosphines were also titrated and the titration curves are shown in Fig. 8. It can be seen that the two derivatives with large substituents (2 and 5, 2-cyanoethyl- and cyclohexyl-) only form stable 2 : 1 complexes. The butyl and phenyl derivatives (3 and 4) form 3 : 1 complexes as well as the 2 : 1. All form

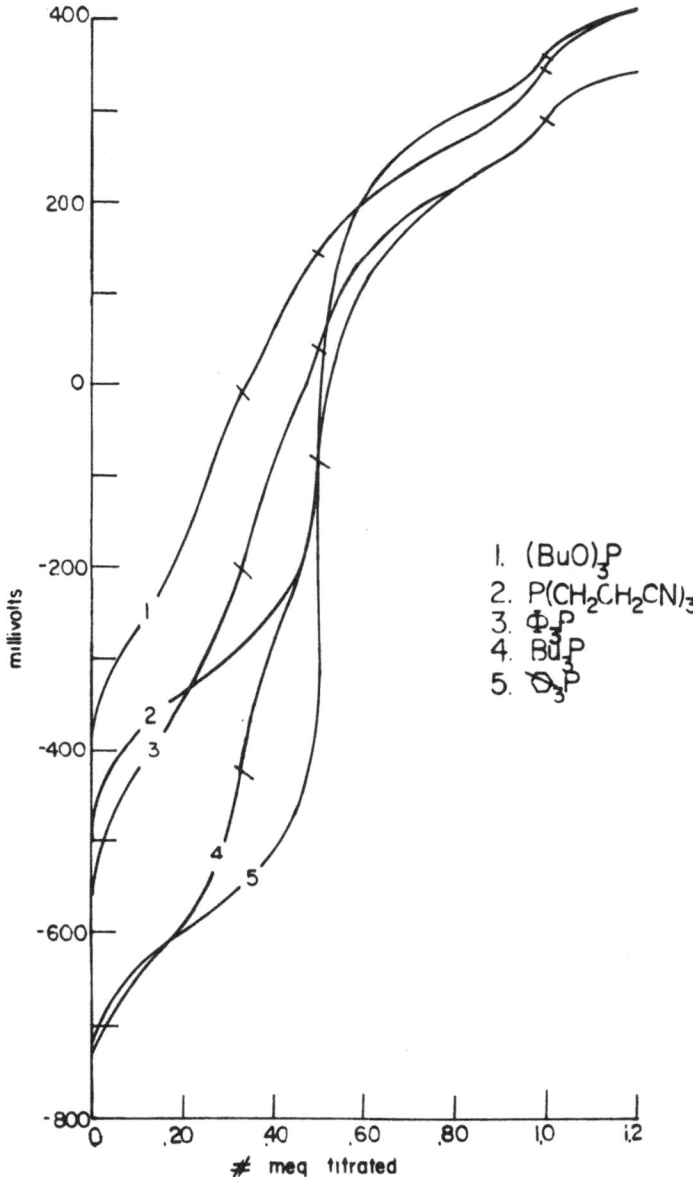

Fig. 8. Titration of substituted phosphines in acetone with silver perchlorate.

weak 1 : 1 complexes. The phosphite derivative, which is also basic to protons, behaves in a manner very similar to the tributylphosphine. The tributylphosphine oxide, however, does not form complexes in acetone which are detectable by this method.

The stability of the complexes again appears to be related to the proton affinity of these molecules and is illustrated in Table III and Fig. 9. Values for the mcp are given for 3 : 1, 2 : 1 and 1 : 1 complexes in the table. Since only some of the phosphines form 3 : 1 complexes it is impossible to make correlation. The 1 : 1 complexes obviously do not correlate with $pK_a(H^+)$ as in the case of the amines. A plot of mcp for the 2 : 1 complexes vs. pK_a is given in Fig. 9. The data are minimal but it appears

Table III. Stability of Silver-Substituted Phosphine Complexes in Acetone

Compound	mcp (mv) AgB_x			$pK_a(H_2O)$
	x = 3	x = 2	x = 1	
$(cy-C_6H_{11})_3P$	–	−575	223	9.70
Bu_3P	−610	−555	278	8.43
ϕ_3P	−372	−303	202	2.73
$(NCCH_2CH_2)_3P$	–	−312	288	1.37
$(BuO)_3P$	−205	−108	248	
Bu_2PH	−334	−236	213	4.51
$(NCCH_2CH_2)_2PH$	−190	−144	265	0.41
$i-BuPH_2$	–	(−41)	(204)	−0.02
$NCCH_2CH_2PH_2$		(−48)	(191)	

that for the tertiary compounds at least a correlation does exist. The relation is different for the secondary and tertiary phosphines, as it is for the similar amines. The greater stability of the complexes for these bases is reflected in the position of the line relative to that of the primary amines, also shown in Fig. 9.

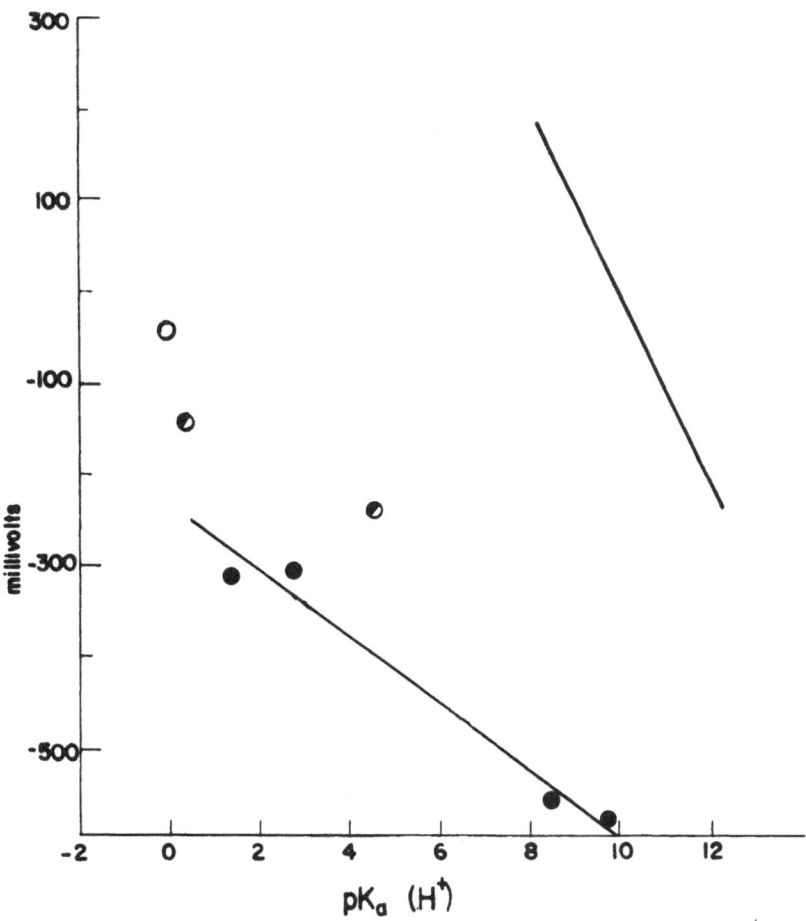

Fig. 9. Stability of $1:2$ silver—phosphine complexes vs pK_a (H^+).

Since both phosphorus and nitrogen compounds are able to coordinate protons as well as silver ion, a question arises as to the stability of the silver complexes in the presence of protons. Data on this point are given on Table IV.

As can be seen, the diphenylguanidine and presumably all of the alkyl amines are strong enough bases to protons to form the N–H bond in preference to the N–Ag. Summation of meq for each quantity gives the total for meq taken. The greater stability of the P–Ag bond and lower proton affinity of the phosphine phosphorus for hydrogen gives a more complex picture for the phosphines. The

Table IV. Effect of Acidity on Formation of Silver–Lewis-Base Complex

Base	mM taken	H^+ added, meq	Ag^+ required, meq	
DϕG	1.264	0.000	0.634	
pK$_a$ 10.0	1.290	0.623	0.324	
	1.036	1.870	0.000	
ϕ_3P	0.696	0.000	0.232	0.345
pK$_a$ 2.73	0.696	0.489	0.232	0.345
	0.696	0.978	0.232	0.345
Bu$_3$P	2.38	0.00	0.77	1.19
pK$_a$ 8.43	2.38	0.98	0.46	0.70
	2.38	2.45	0.00	0.57*

*Potential shift in curve.

triphenyl derivative is unaffected by the hydrogen ion content of the solution at the concentrations studied. The tributyl derivative, which has both a greater proton and silver-ion affinity than the triphenyl derivative, is similar to the amines in its behavior. However, even in an excess of acid there is evidence of complex formation, as is shown in the data presented here and in Fig. 10. The necessity of controlling hydrogen ion concentration for these titrations is obvious.

One final consideration has been given to a molecule that contains both nitrogen and phosphorus functions. Such a molecule is shown in Fig. 11. The molecule is tribasic but when titrated with protons only takes up two equivalents of hydrogen ions. The strongest basic center is one of the nitrogens. The position of the second proton is problematical.

When the same molecule is titrated with silver ions, complexes appear to exist for a 2 : 1 mole : mole compound and also at the 1 : 1 and 1 : 2 points. Six donor molecules are present in the 2 : 1 complex but only a maximum of four can be involved. The 2 : 1 complex may again involve chelation effects with both phosphorus and two nitrogen atoms or may be simply coordination of two phosphorus atoms. The low mcp value would tend to favor chelation, since the mcp value for even tributylphosphine is −555 mv while for this compound it is −621 mv. At the 1 : 1 point a six-membered $-\overset{..}{\underset{|}{P}}-C-C-C-\overset{..}{\underset{|}{N}}-$ ring may be formed. The 1 : 2 complex would indicate silver atoms coordinating with only a single function in the molecule. All of these titration breaks indicate rearrangements of the original molecular complex to a new form.

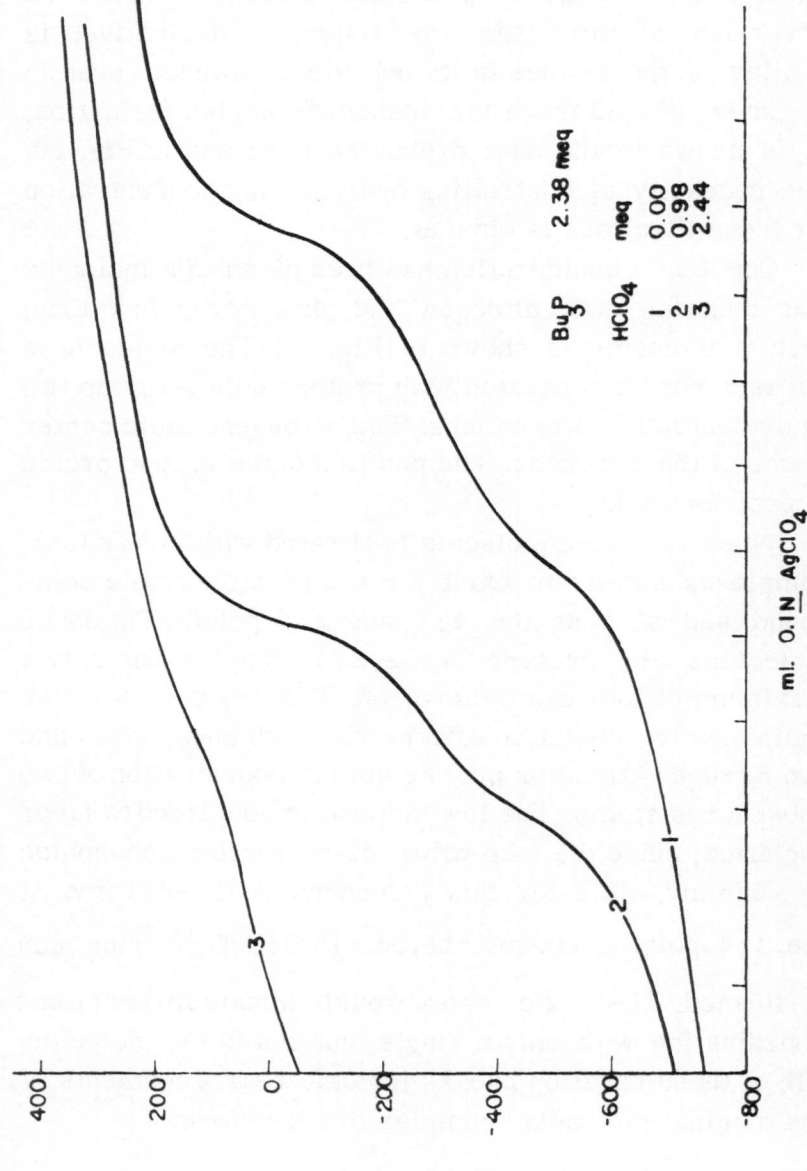

Fig. 10. Effect of acidity on silver–tributylphosphine complexes.

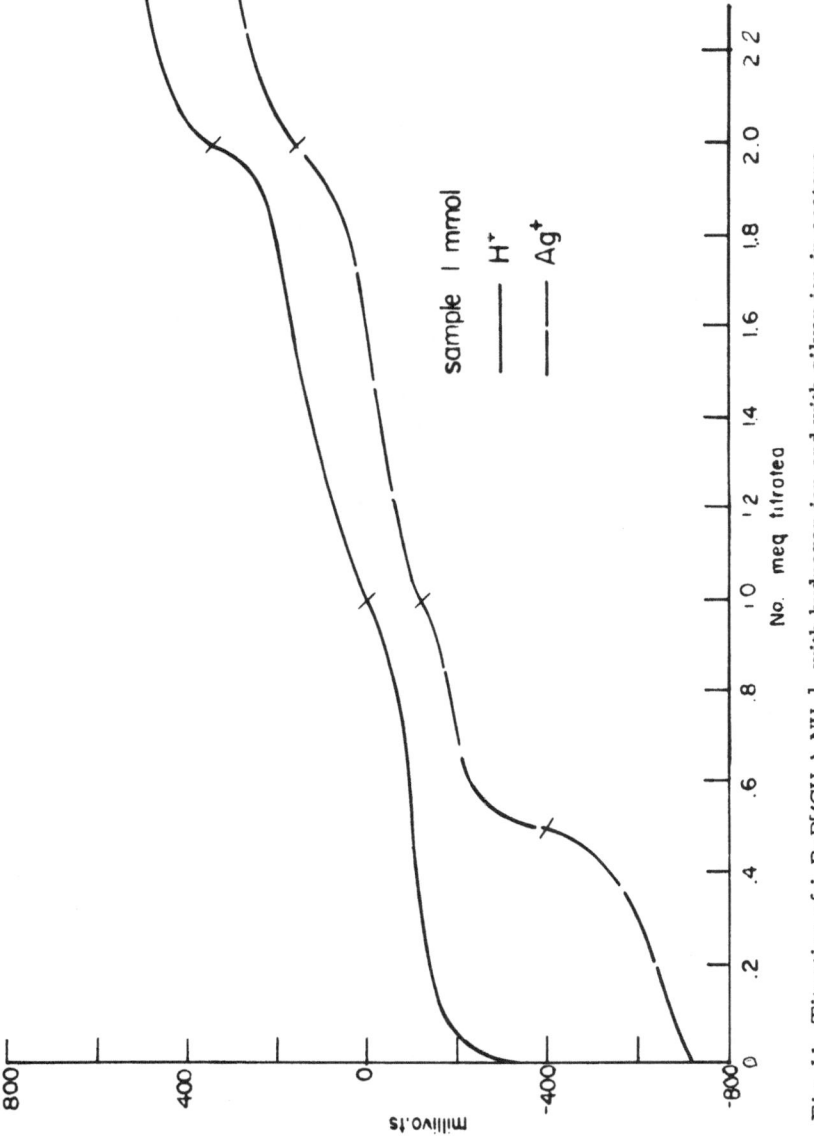

Fig. 11. Titration of i–BuP[(CH$_2$)$_3$NH$_2$]$_2$ with hydrogen ion and with silver ion in acetone.

Bases such as dimethylsulfoxide and acetonitrile cannot be titrated in acetone with silver ion, implying that the complexes are not much more stable than the acetone-silver complex.

In summary, it appears that the formation of silver complexes is a function of proton affinity and ultimately of $\Sigma\sigma^*$ values for both phosphorus and nitrogen compounds. However, the stability of the complex depends on the extent of the substitution. The utility of silver ion for titration of compounds of the phosphine class is potentially much more useful than for the amines. It is probable that silver titrations can be used to differentiate P and N atoms either in mixtures or in a single molecule.

The most important factors influencing the formation of such complexes include the coordinating atom, substitution, type and extent, ability to form chelates, acidity of the solution, and finally, of course, the solvent itself, which limits the electrodes' ability to detect complex formation.

REFERENCES

1. Bruckenstein, S., and Kolthoff, I. M. J. Am. Chem. Soc., Vol. 78, p. 2974, 1956.
2. Hall, H. K., Jr. J. Phys. Chem., Vol. 60, p. 63, 1956.
3. Fritz, J. S., and Yamamura, S. S. Anal. Chem., Vol. 29, p. 1079, 1957.
4. Streuli, C. A., and Miron, R. R. Anal. Chem., Vol. 30, p. 1978, 1958.

ALKOXIDES OF QUATERNARY AMMONIUM BASES. A NEW TYPE OF STRONGLY BASIC TITRANT

Walter R. Heumann

*Chemistry Department, University of Montreal
Montreal, Quebec*

When titrating very weak acids, such as phenols, enols, or sulfonamides, in nonaqueous solvents, the end-point detection is easier and more accurate when stronger bases are used as titrants. This is because both the potential jump in a potentiometric titration and the color change in a visual titration become more pronounced.

The bases most frequently used at present are either the alkoxides of the alkali metals, such as potassium methoxide and sodium ethoxide, or the hydroxides of quaternary ammonium bases, in particular, tetrabutyl-ammonium hydroxide (Bu_4NOH). The Brønsted bases in the first case are the alkoxide anions, CH_3O^- or $C_2H_5O^-$, and in the latter case the hydroxyl anion, OH^-. Although the alkoxide anions are stronger bases than the hydroxyl, in practice Bu_4NOH gives results at least as good as the alkoxides. This may be ascribed to the intrinsic basic strength of the secondary bases from which the anionic Brønsted bases are derived by dissociation. This intrinsic basic strength depends in part on the cation, i.e., on the alkali metal cations and quaternary ammonium cations, respectively, or in other terms on the ease with which these cations release the anions by dissociation. It is well known that in aqueous solutions many quaternary ammonium hydroxides are stronger bases than the hydroxides of the alkali metals.

One should therefore expect that the combination of the stronger basic anion alkoxide with the cation giving the greater intrinsic basic strength, namely, the quaternary ammonium cation, should lead to a titrant base of greater basic strength than that of any one of the abovementioned usual bases. These quaternary ammonium alkoxides are little known and the pertinent literature is almost nonexistent and rather vague, describing them as rather unstable and easily decomposed by both humidity and heat.

We have prepared 0.3 M solutions of tetrabutylammonium ethoxide by adding a 3 M ethanol solution of sodium ethoxide to an approximately 0.3 M solution of tetrabutylammonium bromide in benzene, whereby sodium bromide precipitates and the quaternary ethoxide remains in solution

$$Bu_4N \cdot Br + Na \cdot OC_2H_5 \rightarrow NaBr + Bu_4N \cdot OC_2H_5$$

This method was described about three years ago in a paper on the methylation of morphine, where a quaternary ethoxide is used as the methylating agent [1] and a similar method was described by Croxall et al. in "Organic Syntheses" shortly afterwards [2].

In order to compare the basic strength of this solution of the quaternary ethoxide with that of the corresponding hydroxide and sodium ethoxide solutions, we carried out potentiometric titrations of 0.05 M solutions of phenol with these titrants in about 0.3 M solutions, using protolytic solvents of different basic strength and working under strictly anhydrous conditions.

Before discussing the results, I should like to mention the equipment used. Figure 1 is a schematic sketch of the titration vessel, which consisted of a closed glass cylinder, 60 mm in diameter, having a capacity of 250 ml. It has a conical bottom with a magnetic stirring bar and

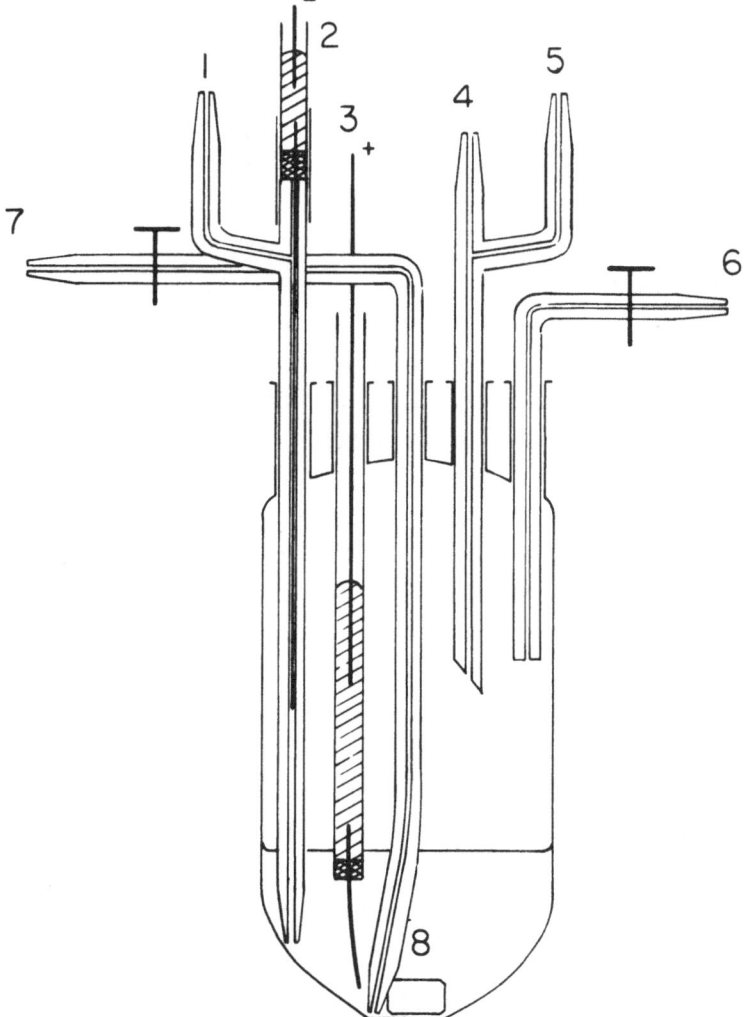

Fig. 1. Titration vessel: (1) titrant solution, (2) reference electrode, (3) indicator electrode, (4) phenol solution, (5) solvent, (6) nitrogen, (7) emptying, and (8) stirring bar.

five necks with ground-glass joints 7/25 for the following connections: (1) titrant solution, (2) reference electrode, (3) indicator electrode, (4) phenol solution to be titrated, (5) solvent for diluting or rinsing, (6) nitrogen for sweeping or producing an inert atmosphere, and (7) emptying. The titrant and phenol solutions were delivered from automatic burets of the Fisher-Porter type which are filled by gravity and thus keep the movement of gases between the atmosphere and the interior at a minimum. The solvent was delivered from an aspirator bottle with a ground-glass bottom outlet. All stopcocks were of Teflon and all ground joints fitted with Teflon sleeves; connections between glass tubes were either thin-walled Teflon tubing or silicone rubber tubing, both of which resist such solvents as aliphatic amines. All connections with the atmosphere went through absorption tubes containing both magnesium perchlorate and soda-lime.

The indicator electrode was an anodized platinum wire while the reference electrode placed in the flow of the titrant was of pure platinum. This electrode system was first suggested by Harlow et al. [3] and showed good reproducibility. We also tried palladium and iridium, either pure or anodized, but the results were less satisfactory. The electrodes were connected to a Beckman model H-2 pH meter. With this setup, series of titrations can be carried out without disconnecting between consecutive titrations; thus, the inert atmosphere can easily be maintained.

The first series of results is shown in Fig. 2, which presents the neutralization curves obtained with ethylenediamine as the solvent when using different titrant bases. The curves are displaced vertically for better readability.

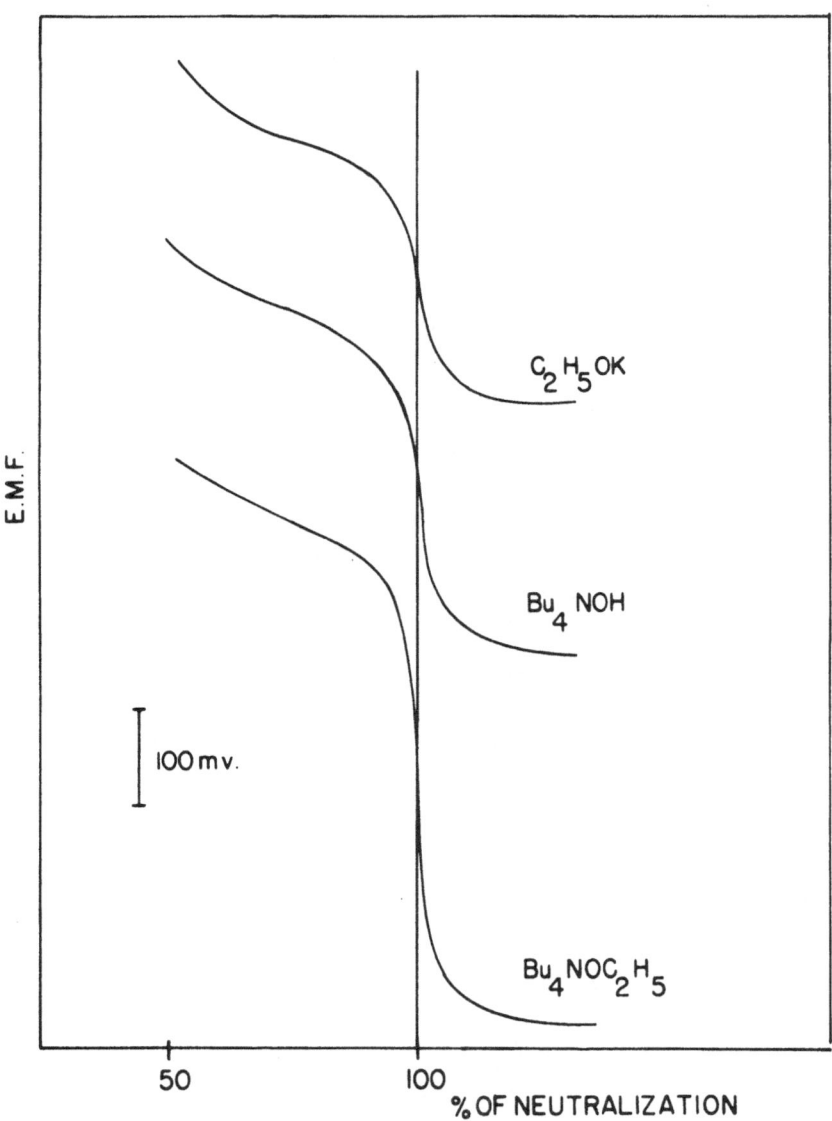

Fig. 2. Titration curves of 0.05N phenol in ethylenediamine with different 0.3N titrant bases.

The difference between sodium ethoxide and tetrabutyl-ammonium hydroxide is not of any importance, the deflections of their respective curves being of the usual size of 250 to 300 mv. The ethoxide of the quaternary base, however, shows a distinctly more-pronounced deflection of 5 to 600 mv. This seems to be evidence for the prediction that the quaternary ammonium alkoxides should be stronger bases than the other types of titrants.

Figure 3 shows the curves we obtained when titrating solutions of phenol with tetrabutylammonium ethoxide in different solvents. As could be expected, the potential change is rather gradual when aliphatic alcohols are the solvents, as these are too weakly basic or are rather amphoteric solvents. More interesting, however, is the difference between n-butylamine and ethylenediamine. Although both are of about the same basic strength, the titration in n-butylamine shows a smaller potential jump than that in ethylenediamine. This is apparently due to the fact that the former is less polar, its dielectric constant being 5.5 as contrasted with 16 for ethylenediamine. The conclusion to be drawn from this seems to be that ethylenediamine is a better solvent than n-butylamine for this type of titration.

It remains to be mentioned that contrary to what we believed together with others, the solutions of quaternary alkoxides appear to be quite stable at room temperature when well protected from carbon dioxide and humidity. We kept solutions of tetrabutylammonium ethoxide or iso-propoxide in ethanol—benzene 1:10 in automatic burets at 25 to 30°C during a rather humid summer season and after two months they were practically unchanged, giving the same titration curves as before.

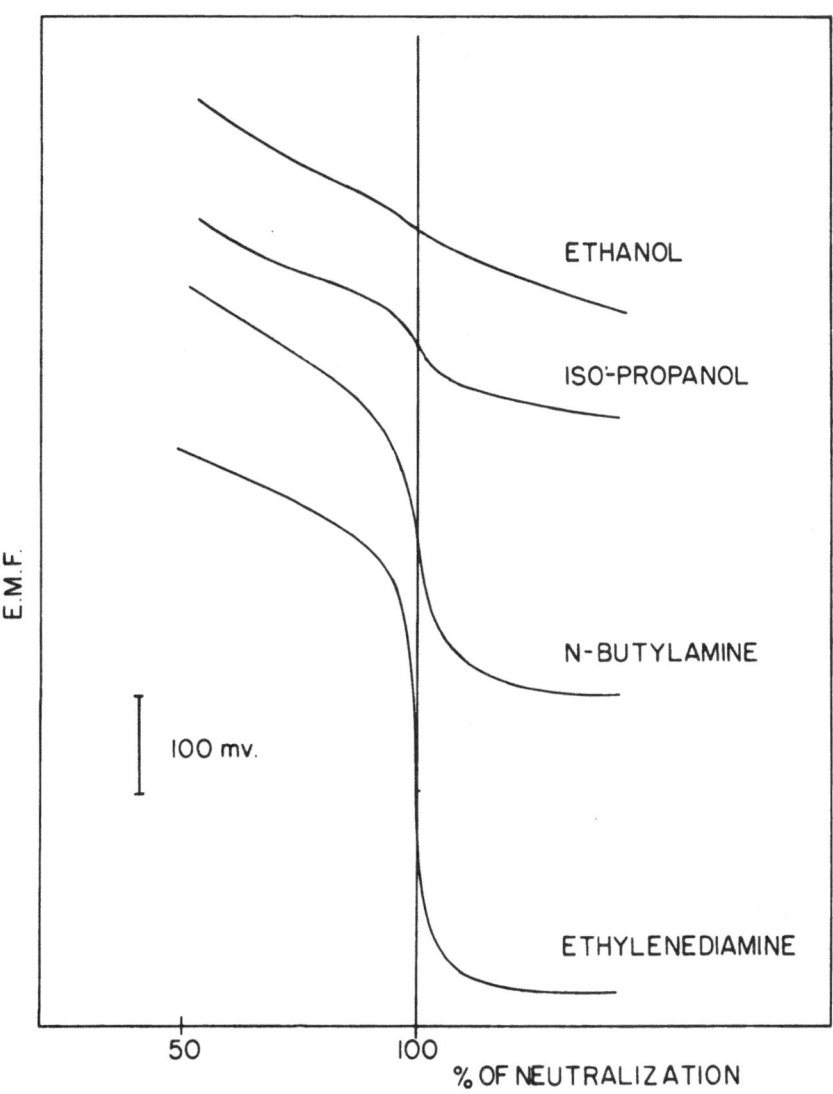

Fig. 3. Titration curves of 0.05N phenol with 0.3N tetrabutylammonium ethoxide in different solvents.

This observation is in accordance with the method of Croxall et al. [2], according to which a crystalline quaternary ethoxide can be prepared by evaporating an ethanol solution to dryness at 40°C.

REFERENCES

1. Heumann, W. R. Bulletin on Narcotics, Vol. X, No. 3, pp. 15-17, 1958.
2. Croxall, W. J., et al. "Organic Syntheses" (John Wiley and Sons, New York, 1958), Vol. 38, pp. 5-8.
3. Harlow, G. A., et al. Anal. Chem., Vol. 28, pp. 784-86, 1956.

APPLICATION OF NONAQUEOUS TITRIMETRY TO PHARMACEUTICAL ANALYSIS

L. G. Chatten*

*Department of National Health and Welfare
Ottawa, Ontario*

As long ago as 1927 when Conant and Hall were carrying out their fundamental studies of the behavior of amines and metal acetates in glacial acetic acid, their observations led them to state that "Much important chemistry has been obscured by our slavish devotion to water." Yet, it was many years following that profound statement before nonaqueous titrimetry was to become a well-established analytical technique, commanding the respect which it enjoys today. It is safe to say that the true value of acid-base titrations in nonaqueous solvents was not fully appreciated until some time after World War II. During the past decade there have been a great number of papers published in many languages which deal with the analytical applications of this technique. Because of its wide applicability, this discussion will be limited to some of our own experiences and I will attempt to give some emphasis to that material which is, as yet, unpublished.

Our aim has been that wherever the existing official or recognized method of assay is a gravimetric determination, an aqueous titration of any type, or a kjeldahl determination we have tried to develop a nonaqueous titrimetric procedure because of its greater speed and accuracy, and, in some instances, specificity.

*Present address: University of Alberta, Edmonton, Alberta.

EPHEDRINE

One of the first products which attracted our interest was the ordinary ephedrine spray. As you know, this drug is presented in its pharmaceutical form in either oily or aqueous media.

The oily product is very simply dealt with. A measured aliquot of the spray is quantitatively transferred to a titration vessel. The solvent is chloroform and the indicator is crystal violet. The titration is carried to the blue color of the indicator with either acetous perchloric acid or perchloric acid in dioxane.

In order to test the utility of this method, we prepared two stock solutions of oily spray according to the formula which was in the National Formulary at that time and then we carried out several replicate determinations on them (Table I).

Table I. Percent Recovery of Ephedrine From Compound Ephedrine Spray N. F.

	Sample I		Sample II	
Operator	A	B	A	B
Number of determinations	6	6	6	6
Mean recovery of ephedrine alkaloid, %	100.40	100.34	100.14	100.27
Standard deviation, %	0.34	0.43	0.12	0.22

For the aqueous sprays, the accurately measured aliquot was made basic, extraction was carried out quantitatively by chloroform, and the titration was performed in the same manner. Stock solutions of this product were made also and their assays are shown in Table II.

One problem, which we thought might be serious, is the fact that many oily sprays are colored by their manufacturers. The question that arose was whether these colors would prevent the use of an indicator by obscuring the end-point change.

We prepared six colored solutions, each with a different dye, and proceeded to titrate them by the prescribed method (Table III).

The color change at the end point was just as sharp and just as readily visible to the eye as it is in colorless preparations. This is one example of extraneous material

Table II. Percent Recovery of Ephedrine From Ephedrine Sulfate Solution N. F.

	Sample I	Sample II	
Operator	A	A	B
Number of determinations	4	4	4
Mean recovery of ephedrine alkaloid, %	100.52	99.84	100.02
Standard deviation, %	0.43	0.47	0.25

Table III. Percent Recovery of Ephedrine From Colored Solutions of Compound Ephedrine Spray N. F.

Dye	Color of solution	Color of solution on addition of indicator	Color change at end point	Recovery, %
Oil Red XO	Orange	Red	Brownish red	100.16 100.00
Yellow OB	Amber	Reddish orange	Brownish red	100.16 100.00
Sudan III	Orange-red	Rose red	Violet	100.33 100.33
Sudan II	Orange	Orange-red	Brownish red	100.00 100.49
Orange SS	Orange	Red	Brownish red	100.16 100.49
Yellow AB	Amber	Orange-red	Brownish red	100.00 100.33

in the pharmaceutical having no undesirable effect upon the applicability of nonaqueous titrimetry. I will have more to say about this later.

In addition, several manufacturers supplied the complete formulation for their product upon our request and each of the ingredients was tested in the solvent system for interference. None of them interfered in any manner.

CODEINE PHOSPHATE IN APC & C

I would like to mention next the combination of acetylsalicylic acid, phenacetin, caffeine, and codeine, commonly called APC & C, of which there are great quantities sold in Canada every year. Despite the fact that these products have been on the market for about 25 years, until as recently as 1954 no satisfactory quantitative method for the determination of the codeine phosphate content was available. At that time two methods were developed in our laboratory in order to cope with these products. Briefly, the procedures are as follows.

M e t h o d I: To a sample of the powdered tablet representing approximately 30 mg of codeine phosphate, add 5 g phenol and 10 ml $CHCl_3$; stir to dissolve the alkaloidal salt, filter through a pledget of cotton, and wash with a small amount of phenol chloroform solution. Add 50 ml of acetonitrile and titrate potentiometrically.

M e t h o d II: To a similar size sample, add a little water, alkalize, and extract the alkaloid quantitatively with chloroform. Wash the combined chloroform extracts, filter, and reduce volume of chloroform to about 10 ml. Add 50 ml CH_3CN and titrate either potentiometrically or visually using chlorophenol red, which changes from yellow to colorless at the end point.

Some of you may wonder why we used a combination of phenol-chloroform for method I. Codeine phosphate has a solubility of 1 : 5000 in chloroform, yet when phenol was added the alkaloidal salt readily dissolved. It has been suggested that the codeine phosphate may undergo complex formation with the phenol and that the complex generated is readily soluble in chloroform. It is interesting to point out that neither a visual nor potentiometric titration could be performed until acetonitrile was added to the solvent system.

Analysis of codeine phosphate in presence of acetylsalicylic acid, phenacetin, and caffeine by method I gave $100.35 \pm 0.64\%$, and by method II gave $99.0 \pm 0.84\%$.

Application to Pharmaceuticals (Table IV)

In general, there were three companies, F, G, and H, whose products could not be assayed by method I. Upon obtaining the formulations of these products, it was noted that gelatin was a common constituent in all of them.

Glasstone, in his text on physical chemistry, stated that when solutions of gelatin are cooled, the particles form into long threads and eventually become interlocked. It is probable that some of the codeine phosphate became trapped in this mesh and the phenol—chloroform system was unable to reach it. Method II gave satisfactory results, since the addition of water reverses the gelation process and the alkaloid was then available for extraction.

This same solvent system of phenol—chloroform—acetonitrile was used for a number of other organic salts, all of which have a low solubility in chloroform alone. The solubility in every instance was excellent and the recoveries satisfactory in the solvent system mentioned (Table V). A miscellaneous group of pharmaceuticals, including tablets of amphetamine sulfate, d-amphetamine sulfate, phenindamine tartrate, and morphine sulfate, to mention just a few, has been analyzed by this solvent system.

In order to further illustrate the general applicability of nonaqueous titrimetry and without going into detail, I would just like to mention that a method for the assay of cholic, desoxycholic, and dehydrocholic acids was developed. Two solvent systems have been utilized, benzene—methanol and N,N-DMF—chloroform. The indicator was thymol blue and the titrant was potassium hydroxide in methanol (Tables VI and VII).

THE BARBITURATES

The barbiturates have come under a great deal of study and some of our observations may be of interest. The standard method of assay for these pharmaceuticals has been and still is gravimetric. There is no doubt that this is the worst possible procedure that could be imagined for these products. Barbiturates readily undergo degradation to substances which possess no therapeutic efficacy, yet by the gravimetric method all decomposition products are included in the assay as though they were the barbiturate (decomposition products of phenobarbital are—monoureide of phenethyl malonic acid $\xrightarrow{-CO_2}$ phenethylacetylurea $\longrightarrow$ urea and an acid).

Procedures have been developed in our laboratory which, although not perfect, certainly are a considerable

Table IV. Recovery of Codeine Phosphate From Commercial APC and C Tablets

Manu-facturer	Labeled per tablet, mg	Recovered per tablet, mg	
		Method I	Method II
A	32.40	33.18 33.44	32.94
B	32.40	30.41 30.05	30.03
C	32.40	34.20 33.69	
D	32.40	31.93 31.61	*
E	32.40	29.35 29.61	29.74
F	32.40	*	28.53 28.98
J	32.40	33.65 33.19	
B	16.20	14.00 14.25	14.70
A	16.20	16.24 16.31	16.13
G	16.20	*	17.10 16.84

Manu-facturer	Labeled per tablet, mg	Recovered per tablet, mg	
		Method I	Method II
E	16.20	*	14.50
			14.42
H	16.20	*	16.60
			16.21
I	16.20	15.33	15.51
		15.53	
A	8.10	7.91	
		8.07	
D	8.00	7.91	7.89
		7.94	
D	8.00	*	7.97
			7.91
G	8.10	*	8.31
			8.45
H	8.10	*	7.15
			7.11
B	8.10	7.76	7.78
		7.76	

*Interfering substances.

Table V. Titration of Pure Salts

Salt	Potentiometric		Visual		Water, % (Karl Fischer)
	Taken, mg	Recovered, mg	Taken, mg	Recovered, mg	
Amphetamine sulfate	52.0	52.5	52.0	52.5	0
	50.6	50.4	36.0	36.0	
	42.0	42.3	36.9	36.6	
Dexamphetamine sulfate	54.4	54.3	42.1	42.3	0
	44.9	44.5	38.8	38.7	
	45.3	44.8	54.4	54.3	
Ephedrine sulfate	43.6	43.4	45.7	45.4	0
	42.4	42.4	45.0	44.6	
	41.7	41.9	45.3	45.0	
Morphine sulfate	88.6	87.4	88.6	87.4	9.03
	84.8	83.8	89.1	89.6	
	87.6	87.0	84.8	83.8	

Codeine sulfate	83.3	84.3	77.5	76.9	6.39
	70.9	71.9	35.5	36.0	
			70.9	71.9	
Butacaine sulfate	37.4	37.4	Indicator	Unsuitable	—
	77.8	77.4			
	76.0	75.9			
Quinine sulfate	63.4	63.6	Indicator	Unsuitable	4.3
	66.8	67.0			
	66.9	67.1			
Cinchonine sulfate	26.2	25.9	Indicator	Unsuitable	2.77
	27.1	26.6			
	27.7	27.8			
Cinchonidine sulfate	25.7	25.5	Indicator	Unsuitable	3.89
	29.7	28.9			
	25.3	24.7			
Strychnine sulfate	93.7	93.7	79.0	78.5	10.5
	83.9	83.5	93.7	93.7	
	87.6	87.1	87.6	87.1	

Table V. (Continued)

Salt	Potentiometric		Visual		Water, % (Karl Fischer)
	Taken, mg	Recovered, mg	Taken, mg	Recovered, mg	
Physostigmine sulfate	66.3	66.5	66.3	66.5	2.3
	74.2	74.7	66.9	67.2	
	65.3	65.0	68.1	68.4	
Physostigmine salicylate	52.0	51.1	52.0	51.7	0.4
	50.1	49.7	50.1	50.1	
	50.8	50.3	50.8	50.7	
Phenindamine tartrate	45.3	44.6	45.6	44.6	
	55.8	54.6	45.3	44.6	
	50.9	49.8	55.8	54.6	
Pilocarpine nitrate	30.3	30.3	Indicator	Unsuitable	—
	36.7	36.9			
	34.6	36.1			

	Previously reported				
Codeine phosphate			43.4	43.6	5.26
			46.2	46.6	
			44.3	44.6	
Morphine acetate	46.8	45.7	48.0	46.9	9.6
	53.5	51.9	53.5	52.3	
	49.2	47.6	46.8	45.7	
Dihydrocodeinone bitartrate	57.1	58.1	64.3	64.6	8.35
	56.4	56.8	66.3	65.8	
	65.6	65.6	65.6	66.0	
Dioxyline phosphate	41.9	41.7	43.3	43.6	2.14
	62.7	62.8	41.9	41.1	
	58.6	58.1	58.6	58.1	

Table VI. A Comparison of the Results Obtained Using Two Titrants With Benzene-Methanol as Solvent and Thymol Blue as the Indicator

Bile acid	Sample size, mg	Recovery of Potassium Methoxide Solution, %	Sample size, mg	Recovery of Potassium Hydroxide in Methanol, %
Cholic	52.0	99.8	55.0	100.2
	51.8	99.4	49.6	100.4
	57.7	99.3	50.6	99.2
	164.0	99.1	154.0	99.5
	150.6	99.3	165.2	99.6
	156.0	100.3	167.0	99.6
Desoxycholic	51.2	99.6	50.2	100.0
	57.6	98.8	50.0	99.6
	50.2	99.2	52.4	99.6
	154.3	98.4	160.2	99.2
	149.0	98.8	142.2	100.2
	156.6	99.0	173.0	99.4
Dehydrocholic	53.2	100.5	59.6	101.3
	50.6	101.8	54.2	100.2
	51.6	101.3	64.6	101.6
	151.0	101.0	155.6	100.4
	168.8	100.6	140.4	100.6
	151.0	100.2	166.0	100.4

Table VII. Comparison of the Percent Recoveries From Three Commercial Samples of Dehydrocholic Acid Tablets Using Three Different Procedures

Tablet	Recovery, %		
	Method A*	Method B†	Method C‡
1	97.0	96.4	95.7
	98.3	96.2	95.7
	97.8	95.6	96.1
2	105.3	103.3	103.6
	104.5	102.5	101.8
	104.6	103.3	102.6
3	98.4	96.8	97.4
	98.1	97.2	98.0
	98.5	97.0	97.4

*Potassium methoxide solution titrant with benzene—methanol solvent and thymol blue indicator.
†Potassium hydroxide in methanol titrant with N,N-dimethylformamide—chloroform solvent and thymol blue indicator. Filtering technique used.
‡N. F. IX procedure (aqueous NaOH titrant, alcohol—water solvent and phenolphthalein).

Table VIII. Titration of U.S.P. Grade Phenobarbitone Using a Chloroform—Methanol Solvent System

Taken, mg	Recovered, mg	Recovery, %
45.3	45.3	100.0
44.1	43.9	99.5
46.2	46.3	100.2
53.3	53.2	99.8
47.6	47.7	100.2

Mean = 99.9 ± 0.3.

improvement over existing official techniques (Table VIII). A 50 : 1 chloroform—methanol solvent system was used and the indicator was thymol blue. The same technique was extended to amobarbital, barbital, cyclobarbital, pentobarbital, and several less common barbitu-

Table IX. Recovery of Phenobarbitone From Commercial Tablets

Product	Recovery, %	
	Nonaqueous	B.P. 1953
Aminophylline and pheno-barbitone	98.6 96.8	*
Aminophylline and pheno-barbitone	92.1 91.8	*
Aminophylline and pheno-barbitone	90.4 90.5	*
Aminophylline and pheno-barbitone	101.9 101.4	*
Phenobarbitone, gr $1\frac{1}{2}$	98.1 98.3	98.5
Phenobarbitone, gr $1\frac{1}{2}$	98.0 98.2	98.3
Phenobarbitone, gr $1\frac{1}{2}$	89.0 89.1	97.9†
Phenobarbitone, gr 1	91.3 90.9	90.5

rates. Table IX shows the recovery of phenobarbitone from commercial preparations (B.P. gravimetric).

In certain instances the official method gave results which were higher than the nonaqueous. All residues from this procedure were titrated in the solvent system

Table IX. (Continued)

Product	Recovery, %	
	Nonaqueous	B.P. 1953
Phenobarbitone, gr 1	94.5 94.4	97.8†
Phenobarbitone, gr $\frac{1}{2}$	97.5 97.7	97.3
Phenobarbitone, gr $\frac{1}{2}$	92.9 92.6	99.1†
Phenobarbitone, gr $\frac{1}{2}$	94.6 94.4	95.0
Phenobarbitone, gr $\frac{1}{4}$	91.1 91.5	91.4
Phenobarbitone, gr $\frac{1}{4}$	96.1 96.6	97.6
Phenobarbitone with sodium pentobarbitone	87.2 87.7	*

*Not official in B.P.
†Overestimated by B.P. procedure.

of choice and in each instance this gave results which agreed with the nonaqueous procedure and thus showed that some inert substance in three products was being weighed as phenobarbital.

Sodium Barbiturates

As with the free acids, the pharmacopoeias have relied on gravimetric analysis as a means of quantitatively assaying the salts. The N.N.R. uses ultraviolet light and this is an improvement, but, unfortunately, the decomposition products cause some interference here too.

Without wishing to go into much detail I would just like to mention that we have recently developed a method for the salts of the barbiturates. In order to take care of some of the degradation products as well as many tablet excipients, an aqueous solution of the weighed sample is prepared and filtration is then carried out before acidification and the subsequent extraction with chloroform. The remainder of the procedure is the same as that for the free acids.

Both gravimetric and uv methods showed overestimation when compared to the nonaqueous procedure. Again, titration of the residues from the gravimetric assay gave values which agreed closely with those of nonaqueous titration (Table X).

PARA-AMINOSALICYLIC ACID AND SODIUM PARA-AMINOSALICYLATE

With regard to p-aminosalicylic acid and its sodium salt, the official method for these drugs is a nitrite titration which suffers from two major disadvantages, the outside indicator and the fact that anything with a

Table X. Results of the Analysis of Pharmaceuticals Containing the Sodium Salts of Various Barbiturates

Manu-facturer	Barbiturate	Recovery, %		Titration of residue, %
		Nonaqueous*	Official	
A	Amobarbital sodium capsules, 200 mg	92.35 ± 0.40	95.66 95.61	91.97 92.19
A	Amobarbital sodium capsules, 65 mg	90.98 ± 0.46	107.85 112.23	90.68 92.25
A	Amobarbital sodium capsules, 194.4 mg	95.53 ± 0.37	106.84 110.08	95.03 96.33
B	Amobarbital sodium capsules, 64.8 mg	89.33 ± 0.75	93.70 100.23	88.05 89.00
C	Amobarbital sodium capsules, 194.4 mg	89.88 ± 0.92	99.74 99.53	91.78 92.17
D	Butabarbital sodium tablets, 32.4 mg	86.70 ± 1.01	92.88† 93.06	

Table X. (Continued)

Manu-facturer	Barbiturate	Recovery, %		Titration of residue, %
		Nonaqueous*	Official	
D	Butabarbital sodium tablets, 32.4 mg	89.92 ± 0.59	90.88† 91.90	
E	Butabarbital sodium tablets, 24.3 mg	90.16 ± 0.60	94.72† 93.48	
E	Butabarbital sodium tablets, 16.0 mg	85.37 ± 0.99	96.89† 98.24	
F	Butabarbital sodium tablets, 32.4 mg	87.48 ± 0.97	95.45† 95.67	
G	Butabarbital sodium tablets, 30 mg	93.35 ± 0.67	101.81† 101.40	
H	Cyclobarbital calcium tablets, 200 mg	89.75 ± 0.40	93.84 95.67	88.68 91.49

I	Pentobarbital sodium	89.84 ± 0.81	‡	—
J	Pentobarbital sodium capsules, 97.2 mg	100.12 ± 1.26	108.15 104.11	94.28 93.82
E	Phenobarbital sodium tablets, 97.2 mg	84.82 ± 0.89	93.33 99.48	83.27 85.26
A	Secobarbital sodium tablets, 50.0 mg	95.63 ± 0.76	‡	—
A	Secobarbital sodium tablets, 97.2 mg	97.31 ± 0.57	‡	—
A	Secobarbital sodium pulvules, 48.6 mg	95.46 ± 0.72	‡	—
C	Secobarbital sodium capsules, 97.2 mg	95.74 ± 0.51	107.63 103.30	95.71 95.44

*Standard deviation calculated on five determinations.
†The official method is ultraviolet. No residue for titration.
‡No satisfactory results could be obtained by the official method.

primary amino group can be diazotized, so meta-amino-phenol, a breakdown product, is titrated as though it were PAS. Nonaqueous procedures are as follows.

    (a) Titration of PAS in acetone: potassium hydroxide in methanol as titrant; thymol blue indicator.

    (b) Titration of NaPAS in anhydrous methanol: per-chloric acid in dioxane as titrant; thymol blue indicator.

Table XI. Comparative Assays of p-Aminosalicylic Acid Tablets

Product	Nonaqueous	Nitrite method
A	98.8	99.3
	98.4	101.8
	99.1	
B	101.7	100.2
	101.1	99.8
	101.1	
C	101.1	103.8
	101.3	102.8
	100.4	
D	99.4	101.2
	99.4	102.3
	99.6	
E	102.6	104.0
	102.0	103.2
	102.4	

Comparative assays of PAS and NaPAS are given in Tables XI and XII respectively.

By the nonaqueous procedures it is possible to titrate not only PAS in the presence of MAP, but also sodium PAS in the presence of MAP, since the latter compound behaves as a weak base in methanol in the presence of sodium PAS. While one would not expect MAP to interfere in the titration of PAS, it was surprising to learn

Table XII. Comparative Assays of Sodium p-Aminosalicylate Tablets

Product	Nonaqueous	Nitrite method
A	93.8	96.7
	93.3	96.3
	93.4	
B	105.6	108.1
	106.5	107.4
	105.9	
C	99.6	98.1
	99.4	98.5
	99.4	
D	103.0	99.2
	102.2	100.0
	102.6	
E	88.0	88.1
	88.6	89.0
	88.5	

that MAP did not interfere with the assay of the sodium salt. Hence, the nonaqueous methods show greater specificity than the U.S.P. procedures.

Watson, Yokoyama, and Pernarowski were able to show that if the MAP content and free acid value were subtracted from the U.S.P. assay for sodium PAS tablets, the values in every instance but one were almost identical with the nonaqueous values for the sodium salt (Tables XIII to XV).

Similarly, if the MAP content, calculated as PAS, is subtracted from the U.S.P. assay, the result in each instance agreed closely with the nonaqueous titration of the free acid.

Table XIII. Results of Nonaqueous and USP Analyses of Crystalline Sodium Aminosalicylate and Aminosalicylic Acid

	Sodium aminosalicylate		Aminosalicylic acid	
	Nonaqueous	U.S.P.	Nonaqueous	U.S.P.
	100.8%	100.0%	100.8%	100.9%
	100.2%	99.7%	99.8%	100.9%
	100.1%	100.5%	99.4%	99.9%
Mean	100.4%	100.1%	100.0%	100.6%
Standard deviation	± 0.4%	± 0.4%	± 0.7%	± 0.6%

Table XIV. Results of the Analysis of Sodium Aminosalicylate Tablets*

Product	U.S.P.	Method of analysis					Corrected U.S.P.	Nonaqueous
		Corrections for U.S.P.†						
		X	X'	Y	Y'			
A	98.3 ± 0.6	4.5	6.2	0.1	0.2		91.9	91.3 ± 0.3
B	100.1 ± 0.1	1.8	2.5	0.0	0.0		97.6	98.0 ± 0.3
C	96.6 ± 1.1	11.3	15.6	2.8	5.4		75.6	79.1 ± 1.6
D	103.2 ± 0.2	1.2	1.7	0.0	0.0		101.5	100.2 ± 0.5

*Results are expressed as percent recovery of label claim. Standard deviations based on triplicate determinations are reported after each relevant value.
†X = free acid calculated as PAS. X' = free acid calculated as Sodium PAS.
Y = MAP content. Y' = MAP content calculated as Sodium PAS. See text for details.

Table XV. Results of the Analysis of Aminosalicylic Acid Tablets*

Product	U.S.P.	Corrections for U.S.P.†		Corrected U.S.P.	Nonaqueous
		Y	Z		
E	95.6 ± 0.5	3.4	4.8	90.8	89.1 ± 0.1
F	100.3 ± 0.8	1.3	1.8	98.5	97.8 ± 0.4
G	101.4 ± 0.4	0.4	0.6	100.8	98.9 ± 0.8

*Results are expressed as percent of label claim. Standard deviations based on triplicate determinations are reported after each relevant value.
*Y = MAP content, Z = MAP content calculated as PAS. See text for details.

PHENOTHIAZINES

Work on the phenothiazine tranquilizers was begun in our laboratory late in 1956 and was carried on over a period of almost two years. At that time no official methods existed for any of these important drugs and even to the present time I think there are only two that have attained official status.

Two methods were developed which worked equally well for both promazine and chloropromazine.

Method I, for Tablets:

The weighed sample of powdered tablet material is stirred with acetone, filtered using quantitative techniques, and titrated with perchloric acid in dioxane, using methyl red as indicator. The end point is from orange to salmon pink.

Method II, for Ampoules:

After alkalizing the measured aliquot, the drug is extracted quantitatively with hexane. Acetone is added and the same titrant and indicator are used as in method I.

Comparisons were made with another published method and agreement was very satisfactory (Table XVI).

This work was continued for each of the following phenothiazine type tranquilizers: proclorperazine, thiopropazate, perphenazine, mepazine, acepromazine, and triflupromazine. Results for pharmaceuticals are indicated in Table XVII.

TETRACYCLINE ANTIBIOTICS

The wide usage and therapeutic importance of the tetracycline antibiotics have resulted in the appearance of many pharmaceutical forms.

Table XVI. Comparative Recoveries of Nonaqueous and Gravimetric Methods

| Form | Nonaqueous method | | | Gravimetric method |
	Number of estimations	Mean recovery, %	Standard deviation	Mean recovery, %
Pure chlorpromazine hydro-chloride by method I	6	101.4	0.130	—
Pure promazine hydrochloride by method I	6	101.05	0.150	—
Pure chlorpromazine hydro-chloride by method II	6	101.3	0.225	—
Chlorpromazine tablets	10	106.02	0.812	105.4
Promazine tablets	5	96.23	0.472	96.15
Chlorpromazine ampoules	5	110.06	0.689	110.8
Promazine ampoules	5	97.92	0.502	100.1
Chlorpromazine suppositories	5	102.2	*	102.2

*Single suppositories analyzed and therefore standard deviation not representative of method itself.

Table XVII. Results of the Analyses of Pharmaceutical Forms

Pharmaceutical form	Brand name	State of active ingredient	Nonaqueous recovery, %	Nonaqueous standard deviation	Gravimetric recovery, %	Control procedure recovery, %
Triflupromazine tablet	Vesprin	Hydrochloride	99.13	0.655	98.20	99.38
Acepromazine tablet	Plegicil	Maleate	97.01	0.590	97.84	108.00
Acepromazine drops	—	Maleate	98.36	—	103.2	111.00
Acepromazine ampoule	—	Maleate	95.23	—	98.16	92.00
Mepazine tablet	Pacatal	Hydrochloride	100.18	0.610	—	98.96
Mepazine ampoule	—	Acetate	102.10	0.150	—	99.04
Prochlorperazine tablet	Stemetil	Dimaleate	96.63	0.680	—	97.01
Prochlorperazine suppository	—	Free base	102.10	0.162	—	—
Perphenazine tablet	Trilafon	Free base	—	—	103.2	102.55
Thiopropazate tablet	Dartal	Dihydrochloride	99.67	1.62	—	98.24

Table XVII. Recovery of Tetracycline Antibiotics in Various Dosage Forms by the Nonaqueous, Microbiological, and Spectrophotometric Methods

Dosage form	Antibiotic	Nonaqueous method	Microbiological method	Spectrophotometric method
Capsule	Aureomycin hydrochloride	96.01 ± 0.41 (5)*	114 (1)	108.3 ± 0.2 (3)
	Achromycin hydrochloride	99.72 ± 0.66 (6)	107 (1)	103.2 ± 1.1 (3)
	Terramycin hydrochloride	102.7 ± 0.75 (8)†	103 (1)	105.9 ± 1.6 (3)†
Tablet	Achromycin hydrochloride	98.96 ± 0.29 (5)	115 (1)	103.2 ± 1.1 (3)
Ointment	Aureomycin hydrochloride	115.5 ± 0.69 (5)	114 (1)	123.2 ± 0.70 (2)
	Achromycin hydrochloride	118.0 ± 0.53 (5)	107 (1)	118.6 ± 1.6 (3)
	Terramycin hydrochloride with poly- myxin B sulfate	106.5 ± 0.24 (5)†	105 (1)	110.5 ± 1.4 †
Intravenous injection	Terramycin hydrochloride	112.4 ± 0.38 (5)	93 (1)	117.7 ± 0.5 (3)
Suppository	Aureomycin hydrochloride	105.8 ± 0.32 (6)	109.2 (1)	111.5 ± 2.3 (6)

*Numerals in parentheses indicate the number of determinations.
†Calculated as free base $C_{22}H_{24}N_2O_9$.

The standard method of analysis for these products is microbiological and gives a rather large deviation between replicates. Several fluorometric and colorimetric procedures have appeared in the literature, but all are time consuming and have only fair accuracy. Nonaqueous titrimetric methods were developed in our laboratory which could be applied successfully to tablets, capsules, ointments, parenterals, and suppositories with an accuracy of about ±0.5%. Comparative results are shown in Table XVIII.

The solvent system was generally composed of nitromethane, formic acid, and benzene, 50:1:5, and the indicator was a mixture of quinaldine red and methylene blue in anhydrous methanol. The end point was signified by a green color.

THE ANTIHISTAMINES

In order to demonstrate further the wide applicability of nonaqueous titrimetry, I would like to mention just one more class of pharmaceuticals for which analytical procedures have been developed, namely, the antihistamines.

By means of either of two solvent systems, two of us were able to analyze successfully 18 antihistamines and the pharmaceutical forms of 12 of these. In addition, there were seven other drugs which could not be assayed in either solvent system. The following general techniques were utilized.

(a) After quantitatively removing the drug from the powdered tablet or capsule by stirring with chloroform and filtering, an equal volume of glacial acetic acid was added prior to titration; the indicator was crystal violet.

Table XIX. Results of Analyses of Pharmaceuticals

No.	Compound	Total weight of tablet, mg	Declared concentration, mg/tablet	Concentration found, mg/tablet	
				Nonaqueous method	U.S.P. method
1	Bromodiphenhydramine hydrochloride	250.0	25.0*	24.4	25.6[†]
2	Diphenhydramine hydrochloride	141.0	25.0*	25.1	25.5
3	Dimenhydrinate	181.7	50.0	50.2	50.8
4	Doxylamine succinate	178.3	25.0	23.5	25.6
5	Methapyrilene hydrochloride	212.0	25.0	25.8	25.8
6	Pyrilamine maleate	120.0	50.0	47.0	47.6

7	Pyrilamine maleate	131.0	50.0	46.5	46.1
8	Tripelennamine hydrochloride	199.2	50.0	44.9	44.6
9	Thonzylamine hydrochloride	99.5	25.0	23.2	23.3
10	Chlorcyclizine hydrochloride	148.0	50.0	49.6	49.8
11	Chlorcyclizine hydrochloride	186.0	50.0	51.0	50.8
12	Phenindamine tartrate	268.6	25.0	24.6	26.0
13	Isothipendyl hydrochloride	126.6	4.0	4.0	4.1†
14	Chlorpheniramine maleate	195.0	4.0	3.6	3.9
15	Piperinhydrinate	105.4	3.0	3.0	‡

*Capsules; all other products in the form of tablets.
†Manufacturer's method of analysis.
‡Insufficient sample for U.S.P. method.

(b) Following the chloroform treatment, an equal volume of acetonitrile was added before beginning the titration; methyl red was the indicator of choice.

Table XIX on pharmaceuticals shows comparative results.

EXCIPIENT INTERFERENCE IN NONAQUEOUS TITRIMETRY

Since today's pharmaceuticals are complex formulations rather than consisting solely of the active constituents, they present a grave challenge to the analyst. Even when all the ingredients are known, all the dexterity of the analyst is frequently required in order to obtain an accurate and correct analysis. Analytical chemists belonging to law enforcement agencies are rarely in the favored position of knowing the complete formulation of a pharmaceutical at the time of assay. It behooves us, therefore, to develop and use procedures which are not subject to interference by diluents, excipients, lubricants, and other components.

Although the U.S.P. has established nonaqueous titrimetry as the official method for many pure drugs and also the B.P. and Ph.I. are showing a trend in that direction, all three compendia have been reluctant to accept this technique for more than a few of the pharmaceutical forms. We in Pharmaceutical Chemistry Section of the Food and Drug Directorate have taken a rather different view of this and I would like to amplify that statement just a little.

We have been particularly interested in developing an application of nonaqueous titrimetry wherever existing methods have a large inherent error for one reason or

another or where the procedure being used is slow, or
cumbersome, etc. I have given several examples of these.
There are also other criteria that we use; one is the
instance where the tablet contains a large amount of
active component in relation to the quantity of excipients.
Particular examples of this are PAS, (already mentioned)
and the sulfonamides. It has been our experience and
observation that if these tablets contain interfering excipients, the amount is not in excess of 1% of the quantity
of active constituent. We are prepared, therefore, to
accept an error of 1% due to any interfering excipients
in such instances. When one considers that the inherent
error of the official methods for these same drugs ex-
ceeds 5% and may in some instances reach as high as
10%, our decision seems justified.

I do not wish to minimize our concern about the effect
of excipients, particularly with respect to those products
in which the active constituent may range from only 5 or
10 mg to, say, 100 mg, nor do I wish to belittle the
pharmacopoeias for taking such a cautious view toward
this problem. I would like to point out that we have tried,
wherever possible, to circumvent the difficulty by the
careful choice of solvents. For instance, one wouldn't
choose glacial acetic acid for the purpose of extracting
the active component from a tablet or capsule, because
of its wide solubilizing properties and leveling ability.

Saturated solutions were prepared for 27 tablet excipients and titrations were performed to determine the
amount of titrant consumed in each instance. Table XX
syows a comparison between glacial acetic acid and
chloroform.

It is apparent that, while possessing some imper-
fections, chloroform is a much more desirable solvent

Table XX. The Extent of Interference Caused by Commonly Used Excipients

Excipient	0.05 N $HClO_4$, ml	
	$CHCl_3$*	CH_3COOH†
Acacia	0.04	0.15
Beeswax	0.00	0.00
Calcium carbonate	0.00	0.13
Calcium phosphate, dibasic	0.04	2.00
Calcium phosphate, tribasic	0.00	0.58
Calcium stearate	0.06	2 +
Calcium sulfate	0.00	0.70
Carnauba wax	0.00	2 +
Cetyl alcohol	0.02	0.00
Gelatin	0.00	2 +
Lactose	0.00	0.02
Magnesium carbonate	0.00	2 +
Magnesium hydroxide	1.06	2 +
Magnesium stearate	0.14	2 +
Magnesium sulfate	0.00	0.08
Methylcellulose	0.00	0.34
Polyethylene glycol "4000"	2 +	0.06
Polyvinyl pyrrolidone	2 +	2 +
Sodium alginate	0.00	0.98
Sodium benzoate	0.00	2 +
Sodium carboxymethylcellulose	0.06	2 +
Sodium silicate	0.00	2 +
Sodium sulfate	0.00	1.18
Starch (potato)	0.00	0.06
Sugar (powdered)	0.00	0.00
Talcum	0.14	0.50
Tragacanth	0.00	0.38

*50 ml chloroform saturated with excipient.
†50 ml glacial acetic acid saturated with excipient.

for this purpose. Let us look at Table XXI. Magnesium stearate is a common component of tablet formulations and, as you will recall, a saturated solution consumed 0.14 ml of titrant. Yet, when this excipient was added to a solution of pure diphenhydramine hydrochloride, it had no effect upon the percentage of drug recovered. Thus, the diphenhydramine is a stronger base in that solvent system than is magnesium stearate.

A continuation of this work is being carried on at present in our section and there remains much to do.

Tables XXII and XXIII give an idea of the results obtained by titrating saturated solutions of these excipients in a number of solvent systems. Where the solvent system is the most promising, we are in the process of seeing whether those excipients which consume titrant when titrated by themselves will, in actual fact, interfere when a strongly basic or acidic component is present in approximately the proportion in which they occur in pharmaceuticals.

Table XXI. Analysis of Synthetic Tablet Formulation

Added, mg		Active ingredient recovered, %
Active ingredient	Magnesium stearate	
71.00	0	100.25
62.95	0	99.70
62.00	14.45	99.10
66.90	12.20	100.10

Table XXII. Titration of Excipients With Perchloric Acid in 0.05 N Dioxane

Excipients	Solvents											
	Acetone	Acetonitrile	Benzene	Chlorobenzene	Dimethyl-formamide	Ethylene glycol	Hexane	Isopropyl alcohol	Methanol	Nitromethane	Phenol–chloroform–acetonitrile	Propylene glycol
Acacia	0.01	—	—	—	—	—	—	—	0.01	—	—	—
Beeswax	0.01	—	—	—	—	—	—	—	—	—	—	—
Calcium carbonate	—	—	—	—	—	—	—	—	—	—	—	—
Calcium phosphate	—	—	—	—	—	—	—	—	—	—	—	—
Calcium phosphate, tribasic	—	—	—	—	—	—	—	—	—	—	—	—
Calcium sulfate	—	—	—	—	—	—	—	—	—	—	—	—
Carnauba wax	—	—	—	—	—	—	—	—	—	—	—	—
Cetyl alcohol	0.01	—	—	—	—	—	—	—	—	—	—	—
Gelatin	—	—	—	—	—	—	—	—	—	—	—	—
Lactose	—	—	—	—	—	—	—	—	—	—	—	—
Magnesium carbonate	—	—	—	—	—	—	—	—	—	—	—	—

Excipient												
Magnesium hydroxide	–	–	–	–	–	–	–	–	0.30	–	–	–
Magnesium stearate	–	–	–	–	–	2*	–	–	2*	–	–	2*
Magnesium sulfate	–	–	–	–	–	–	–	–	2*	–	–	1⁻
Methylcellulose	–	–	–	–	–	–	–	–	–	–	0.04	–
Polyethylene glycol 4000	–	–	0.4	2*	–	–	–	–	–	1*	–	–
Polyvinyl pyrrolidone	0.33	1.5*	0.08	1*	–	–	–	0.04	0.46	2*	–	0.02
Sodium alginate	–	0.06	–	–	–	–	–	–	0.02	0.05	–	0.04*
Sodium benzoate	0.13	–	–	–	0.16	2*	–	2*	2	0.10	0.62	2*
Sodium carboxy-methylcellulose	0.01	–	–	–	–	–	–	–	0.08	0.05	–	–
Sodium silicate	–	–	–	–	–	2*	–	–	2	0.05	–	2*
Sodium sulfate	–	–	–	–	–	–	–	–	0.17	0.03	–	0.2
Potato starch	0.10	–	–	–	–	–	–	–	–	–	–	–
Sucrose	0.03	–	–	–	–	–	–	–	–	–	–	–
Talcum	–	–	–	–	–	–	–	–	–	–	–	–
Tragacanth	–	–	–	–	–	–	–	–	0.10	–	–	–

– Indicates excipient was not titratable in that solvent.
* Endpoint was not reached at volume of titrant listed.

Table XXIII. Titration of Excipients With Potassium Hydroxide in 0.1 N Methanol

Excipients	Acetone	Acetonitrile	Benzene	Chlorobenzene	Chloroform	Dimethyl-formamide	Ethylene glycol	Hexane	Isopropyl alcohol	Methanol	Nitromethane	Propylene glycol
Acacia	—	—	—	—	—	—	—	—	—	—	—	—
Beeswax	0.4	—	2*	1.04	2*	0.04	—	0.8	0.1	—	—	—
Calcium carbonate	—	—	—	—	—	—	—	—	—	—	—	—
Calcium phosphate	—	—	—	—	—	—	—	—	—	—	—	—
Calcium phosphate, tribasic	—	—	—	—	—	—	—	—	—	—	—	—
Calcium sulfate	—	—	—	—	—	—	—	—	—	—	—	—
Carnauba wax	—	—	—	—	—	0.10	—	—	—	—	—	—
Cetyl alcohol	—	0.03	—	—	—	—	—	—	—	—	—	0.07
Gelatin	—	—	—	—	—	—	—	—	—	—	—	—
Lactose	—	—	—	—	—	—	—	—	—	—	—	—
Magnesium carbonate	—	—	—	—	—	—	—	—	—	—	—	—

Excipient	1	2	3	4	5	6	7	8	9	10	11
Magnesium hydroxide	–	–	–	–	–	–	–	–	–	–	–
Magnesium stearate	–	–	–	–	–	–	0.14	0.03	–	–	–
Magnesium sulfate	–	–	–	–	–	–	2*	–	–	–	–
Methylcellulose	–	–	–	–	–	–	–	–	–	–	–
Polyethylene glycol 4000	–	–	–	–	–	–	–	–	–	–	–
Polyvinyl pyrrolidone	0.15	–	0.09	0.24	–	–	0.2	0.12	–	0.30	–
Sodium alginate	2*	–	–	–	–	–	–	–	–	–	–
Sodium benzoate	–	–	–	–	–	–	0.08	–	–	–	0.03
Sodium carboxy-methylcellulose	–	–	–	–	–	–	–	–	–	–	–
Sodium silicate	–	–	–	–	–	–	–	–	–	–	–
Sodium sulfate	–	–	–	–	–	–	–	–	–	–	–
Potato starch	–	–	–	–	–	–	–	–	–	–	–
Stearic acid	2*	2*	2*	2*	2*	2*	2*	2*	2*	2*	2*
Sucrose	–	–	–	–	–	–	–	–	–	–	–
Talcum	–	–	–	–	–	–	–	–	–	–	–
Tragacanth	–	–	–	–	–	–	–	–	–	–	–

– Indicates excipient was not titratable in that solvent.
*Endpoint was not reached at volume of titrant listed.

When this aspect is finished, we plan to extend our investigations to four or five of the most promising solvents in order to establish the actual scope of their usefulness.

In conclusion, I would just like to state that it has long been the feeling of some critics that the effect which excipients might exert on the quantitative analysis of tablets and capsules by nonaqueous titrimetry is probably so great as to limit its usefulness to the assay of the basic materials alone, or at best, to those manufacturers where the control laboratory might be able to exert its influence in the formulation. From our experiences and in our opinion, such a situation is far from actual fact. We believe that nonaqueous titrimetry will be utilized more and more by control laboratories and official compendia once they fully realize that careful choice of solvent systems can greatly minimize and even eliminate the possibility of interference by excipients.

A RAPID DETERMINATION OF MALEIC ACID AND MALEIC ANHYDRIDE IN MIXTURES OF THE TWO

Louis S. Hurst

The Dow Chemical Co.
Midland, Michigan

ABSTRACT

A differentiating potentiometric titration is used to determine the concentrations of maleic acid and maleic anhydride in a wide range of mixtures of the two. However, the determination of small amounts of the acid in the anhydride is apparently the most useful. In particular when maleic anhydride is used in a Diels-Alder type reaction, contamination with maleic acid may cause rearrangement and polymerization of the diene component.

The titration is made with sodium methylate in methanol using methanol as the sample solvent. Two inflection points are obtained when the reference pH values are plotted against milliliters of titrant. The difference between the inflection points corresponds to one equivalent per mole of maleic acid. In order to make the inflection points more distinguishable for less than 1% of acid, a known amount of acid is added. The total acidity, corrected for the amount of free acid, gives the amount of maleic anhydride. The titration curves are rather characteristic for maleic anhydride and maleic acid.

INTRODUCTION

An analysis was needed which would indicate the purity and acid content of maleic anhydride. Small amounts of acid can affect the copolymerization behavior of the anhydride [1] and may cause rearrangement and polymerization of the diene component in Diels-Alder reactions [2].

Standardized solutions of tertiary amines such as tripropylamine and N-ethylpiperidine have been used by Siggia and Floramo [1] to titrate maleic and phthalic acids in their anhydrides. This method is rather specific for maleic acid since the acids which might be included in the determination are those with dissociation constants of 10^{-3}, or greater, in water. This would not include acids such as succinic (dissociation constant, $7 \cdot 10^{-5}$) or fumaric ($9.3 \cdot 10^{-4}$).

Anhydrides have been determined in the presence of their acids by reaction with excess aniline and determination of the excess [3]. The acid content is then found by difference in anhydride and total acidity by hydrolysis.

Mixtures of anhydride and acid have been determined by difference using either sodium methylate in methanol to titrate the free acid plus one equivalent per mole of the anhydride and sodium hydroxide [4], or trimethyl-benzylammonium hydroxide [5] in the presence of water to titrate the free acid plus two equivalents per mole of the anhydride. The determination of free acid by the difference in the two titrations yields poor results for samples of low acid content.

EXPERIMENTAL

Potentiometric titration of maleic anhydride in water with 1N sodium hydroxide converts the anhydride to the dibasic salt; this is probably the most accurate way to determine the total acidity. Since the first carboxyl (dissociation constant, $1.4 \cdot 10^{-2}$) produces an inflection point, any acid impurity with a dissociation constant less than 10^{-3} titrates after the first inflection point. The titration of maleic anhydride in absolute methanol with sodium methylate produces a single symmetrical inflection curve (Fig. 1). Small amounts of maleic acid or similar di-

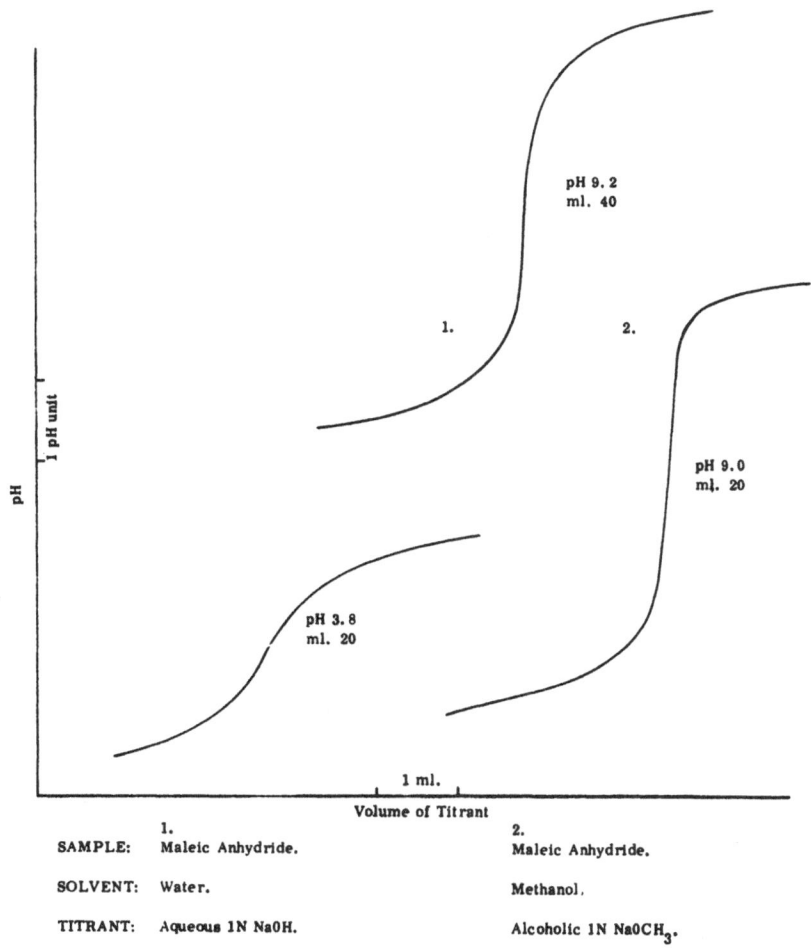

	1.	2.
SAMPLE:	Maleic Anhydride.	Maleic Anhydride.
SOLVENT:	Water.	Methanol.
TITRANT:	Aqueous 1N NaOH.	Alcoholic 1N NaOCH$_3$.

Fig. 1

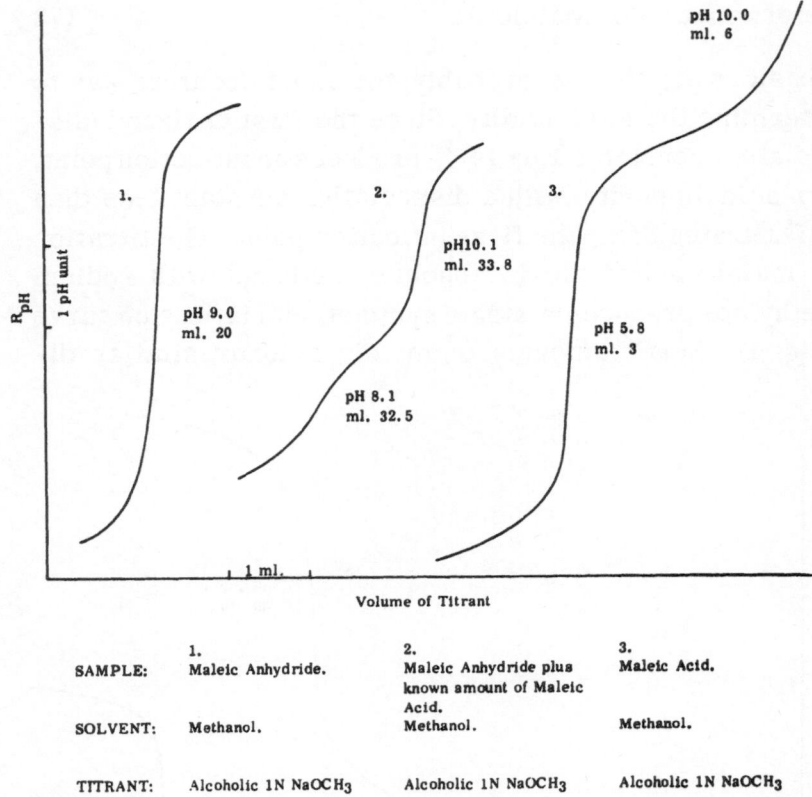

Fig. 2

	1.	2.	3.
SAMPLE:	Maleic Anhydride.	Maleic Anhydride plus known amount of Maleic Acid.	Maleic Acid.
SOLVENT:	Methanol.	Methanol.	Methanol.
TITRANT:	Alcoholic 1N NaOCH₃	Alcoholic 1N NaOCH₃	Alcoholic 1N NaOCH₃

basic acids produce two inflections of approximately equal height (Fig. 2). When 1.28 meq of acid was added, 1.30 meq was recovered, the difference being the amount of acid originally present in the anhydride, 0.11%. The titration of maleic acid in methanol shows the first inflection curve is much sharper than the second. Four commercial samples were shown to contain from 0 to 4% acid when titrated with 1N sodium methylate. Since the inflection curve for less than 1% acid is not well defined, a known amount of acid is added and the titrant is switched from 1N to 0.1N at the approach to the first inflection point (Fig. 3). Percent composition is shown in Table I.

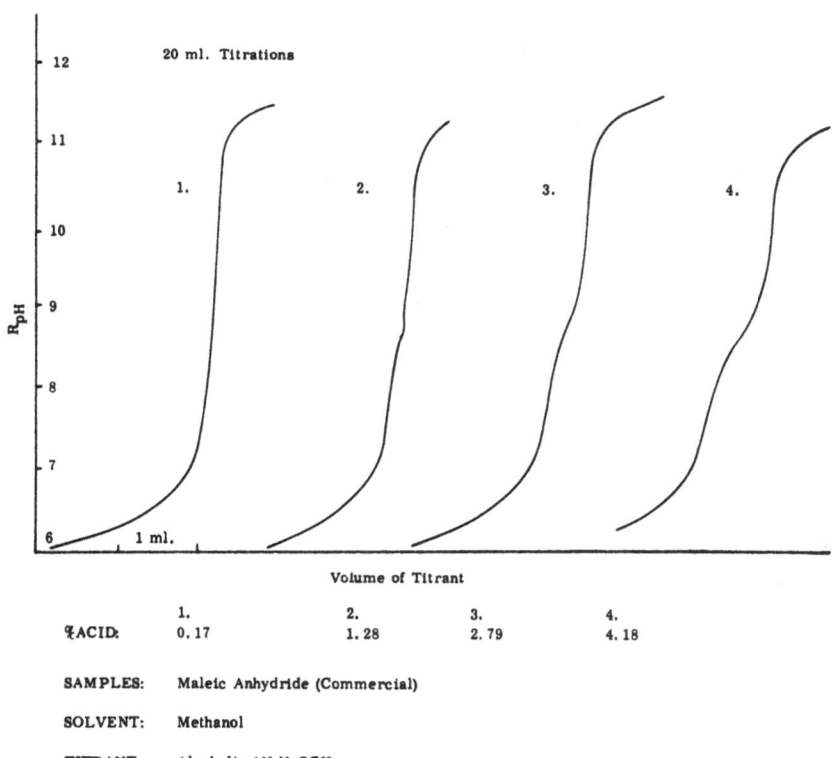

Fig. 3

Table I

No.	Maleic acid by NaOCH$_3$, %	Maleic anhydride by NaOH difference, %	Maleic anhydride by NaOCH$_3$, %
1	0.17*	99.75	99.88
2	1.28*	98.70	98.76
3	2.95*	97.03	96.91
4	4.18*	95.90	95.95
5	28.60†		70.95

*Commercial samples.
†99.1% acid used to make up this sample.

When 40.4 mg of maleic acid was added to 4 g of the anhydride, 47.1 mg was recovered, indicating the sample contained 0.17% maleic acid (Fig. 4). Recovery of maleic acid from a 28.60% acid in anhydride mixture was 100.1%.

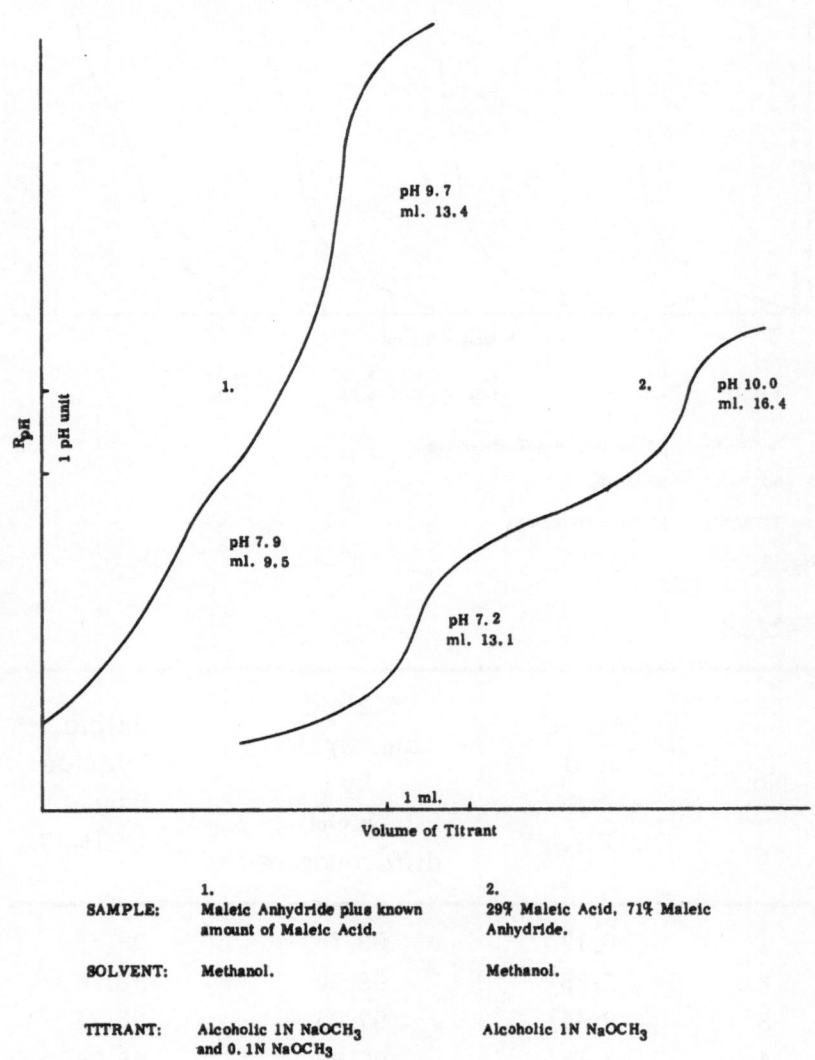

SAMPLE:

SOLVENT:

TITRANT:

	1.	2.
SAMPLE:	Maleic Anhydride plus known amount of Maleic Acid.	29% Maleic Acid, 71% Maleic Anhydride.
SOLVENT:	Methanol.	Methanol.
TITRANT:	Alcoholic 1N NaOCH₃ and 0.1N NaOCH₃	Alcoholic 1N NaOCH₃

Fig. 4

Apparatus

pH meter, Leeds and Northrup, Cat. No. 7664, or equivalent.

Reagents

Absolute methanol of very low water content.

Sodium methylate, 1N solution. Dissolve 54 g of the reagent in methanol and dilute to 1 liter with the same. Standardize against benzoic acid or other suitable standard.

Sodium methylate, 0.1N solution. Dilute 50 ml of the 1N reagent to 500 ml with absolute methanol.

Maleic acid solution. Dissolve 0.400 g of maleic acid in absolute methanol and dilute to 100 ml with the same.

PROCEDURE

Weigh accurately 2-4 g (use 4 g if the acid content is less than 2%) in a 150-ml beaker and dissolve with 40 ml of absolute methanol. Titrate with standard 1N sodium methylate. If the first titration reveals less than 1% acid, then a 10-ml aliquot of the maleic acid solution is added to another sample and the titrant is switched from 1N to 0.1N solution at the approach to the first inflection.

CALCULATIONS

One equivalent per mole of maleic acid is titrated between the inflection points.

A = ml of titrant between inflection points × normality

B = meq of acid added.

$$\% \text{ Maleic acid} = \frac{(A-B)\ 11.61}{g \text{ of sample}}$$

One equivalent per mole of the anhydride and acid is titrated at the first inflection point.

$C = A - B$

$D =$ ml of titrant to first inflection point $\times$ normality

$$\% \text{ Maleic anhydride} = \frac{(D-C)\ 9.806}{g \text{ of sample}}$$

RESULTS

Maleic anhydride calculated from water and methanol titrations were in agreement by 0.15%. The sum of the components reveals a precision of ±0.1%. Precision in determining the acid is ±0.02 to ±0.05%.

CONCLUSIONS

Dibasic acids such as fumaric and succinic would be included in the determination of the acid. If other acids are present, the sodium hydroxide titration would probably indicate their presence and would also serve as a check for the anhydride content.

Determination of the unsaturation may also be used to prove the identity of the components.

REFERENCES

1. Siggia, S., and Floramo, N. A. Anal. Chem., Vol. 25, p. 797, 1953.
2. Royals, E. E. "Advanced Organic Chemistry" (Prentice-Hall Inc., Englewood Cliffs, N. J.), 2nd printing, p. 410.
3. Siggia, S., and Hanna, J. G. Anal. Chem., Vol. 23, p. 1717, 1951.
4. Huhn, H., and Jenckel, E. Z. Anal. Chem., Vol. 163, p. 427, 1958.
5. Patchornik, A., and Rogoqinski, S. E. Anal. Chem., Vol. 31, p. 985, 1959.

TITRATIONS OF 1,4-DIPHENYLPIPERAZINE

Louis S. Hurst

The Dow Chemical Co.
Midland, Michigan

ABSTRACT

1,4-Diphenylpiperazine can be titrated in acetic an-hydride with perchloric acid. The titration curve resulting from plotting millivolts vs. volume of titrant gives two equal and well-defined inflection points. Diphenylpiper-azine is a by-product in the synthesis of diphenylethylene-diamine and may be determined in the presence of the same by acetylation and titration in acetic acid with perchloric acid. It may be identified by separation and titration, obtaining the very characteristic curve in acetic anhydride. Interferences from other tertiary amines and aniline are dealt with in these determina-tions of diphenylpiperazine.

INTRODUCTION

1,4-Diphenylpiperazine is found as an impurity in N,N'-diphenylethylenediamine, which is used as an anti-oxidant in rubber and as a reagent in the quantitative precipitation of water-soluble aldehydes [1].

Potentiometric titration of 1,4-diphenylpiperazine in a solution of acetic acid and anhydride with 0.1N per-chloric acid has been reported along with potentiometric titrations of other 1,4-di substituted piperazines widely used as pharmaceuticals [2].

EXPERIMENTAL

Titration of a pure sample in acetic anhydride with 0.1N perchloric acid gives two inflection points of equal magnitude (Fig. 1). Purity of the sample is established

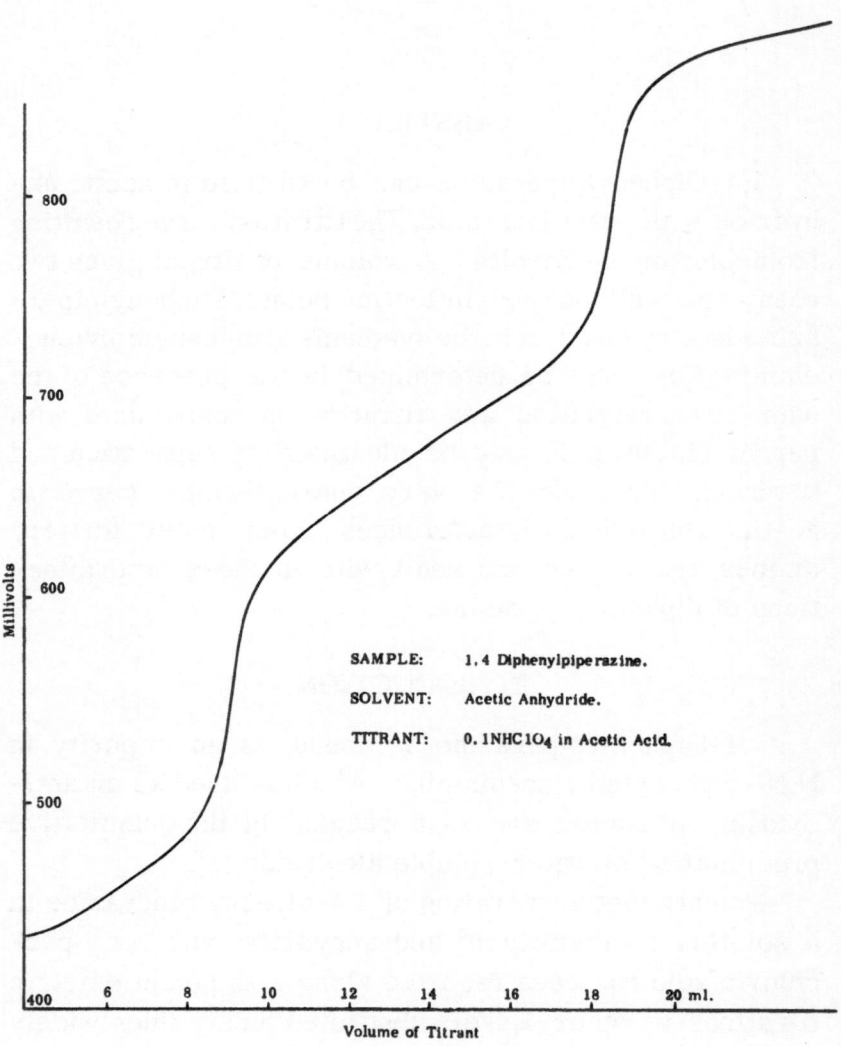

SAMPLE: 1.4 Diphenylpiperazine.

SOLVENT: Acetic Anhydride.

TITRANT: 0.1NHClO$_4$ in Acetic Acid.

Fig. 1

by comparison of the volumes of titrant used to reach the first and second inflections, the first volume being half of the second for high-purity material.

1,4-Diphenylpiperazine can be determined in N,N'-diphenylethylenediamine by acetylating the secondary amine and titrating with $0.1N$ perchloric acid in acetic acid. One equivalent per mole of the tertiary amine is titrated in this case. In this commercial sample 13.75% 1,4-diphenylpiperazine was found. Total alkalinity was best determined by adding a known excess of perchloric acid and back-titrating with sodium acetate (Fig. 2). The two components calculated from these results totaled 99.7%. No aniline or other impurities could be detected. Isolation of the 1,4-diphenylpiperazine was performed by precipitating and filtering the N,N'-diphenylethylene-diamine from an acidic water solution with formaldehyde. The 1,4-diphenylpiperazine was precipitated and filtered by neutralizing the solution. Upon drying and titrating in acetic anhydride, the compound gave the characteristic curve obtained by titrating the pure 1,4-diphenylpiper-azine (Fig. 1).

Other impurities in commercial samples of N,N'-diphenylethylenediamine are indicated by a direct titra-tion in acetic acid (Fig. 3). Small amounts of aniline are present; however, this does not account for the difference in inflection points. N,N'-diphenylethylenediamine gives two inflection points when titrated in acetic acid, the first being very weak. Tertiary amine determinations in acetic anhydride would include other tertiary amines if present; therefore, separations must be performed before titrating the 1,4-diphenylethylenediamine.

Precipitation of the sample in water by neutralizing an acidic solution will separate water-soluble amines such as aniline. Precipitation of the N,N'-diphenylethyl-enediamine and recovery of the 1,4-diphenylpiperazine, as indicated in the analysis of the commercial sample, produces a sample when titrated in acetic anhydride that gives three inflection points. The first is an impurity and the next two are characteristic of 1,4-diphenyl-piperazine.

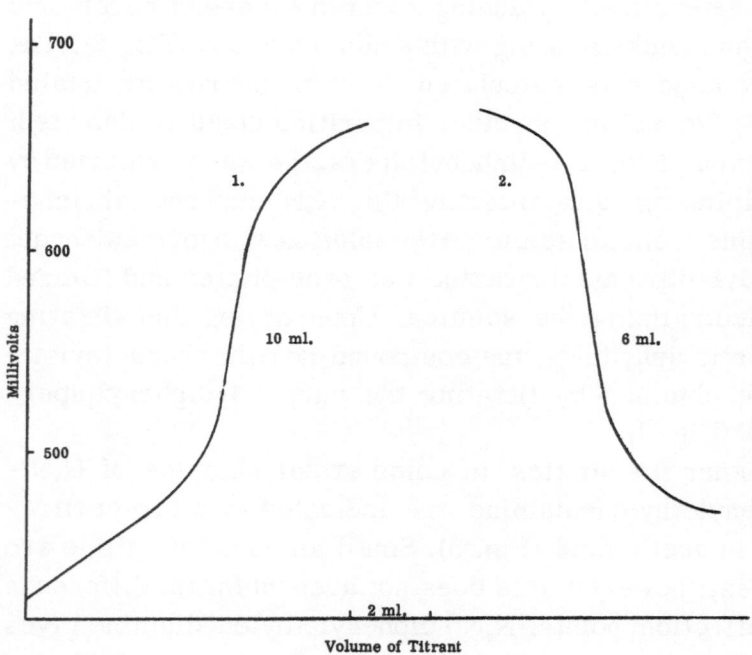

SAMPLES: N, N' Diphenylethylenediamine(Melt. Pt. 65-67° C.) c.p.

	1.	2.
PREPARATION:	Acetylated by heating with Acetic Anhydride.	Known excess of $HClO_4$ added.
SOLVENT:	Acetic Anhydride.	Acetic Acid.
TITRANT:	0.1N $HClO_4$ in Acetic Acid.	0.1N NaOAc in Acetic Acid.

Fig. 2

Apparatus

pH meter, Leeds and Northrup, Cat. No. 7664 or equivalent, equipped with glass and calomel electrodes.

Reagents

Perchloric acid, 0.1 N solution. Dilute 10 ml of the 72% perchloric acid to 1 liter with glacial acetic acid.

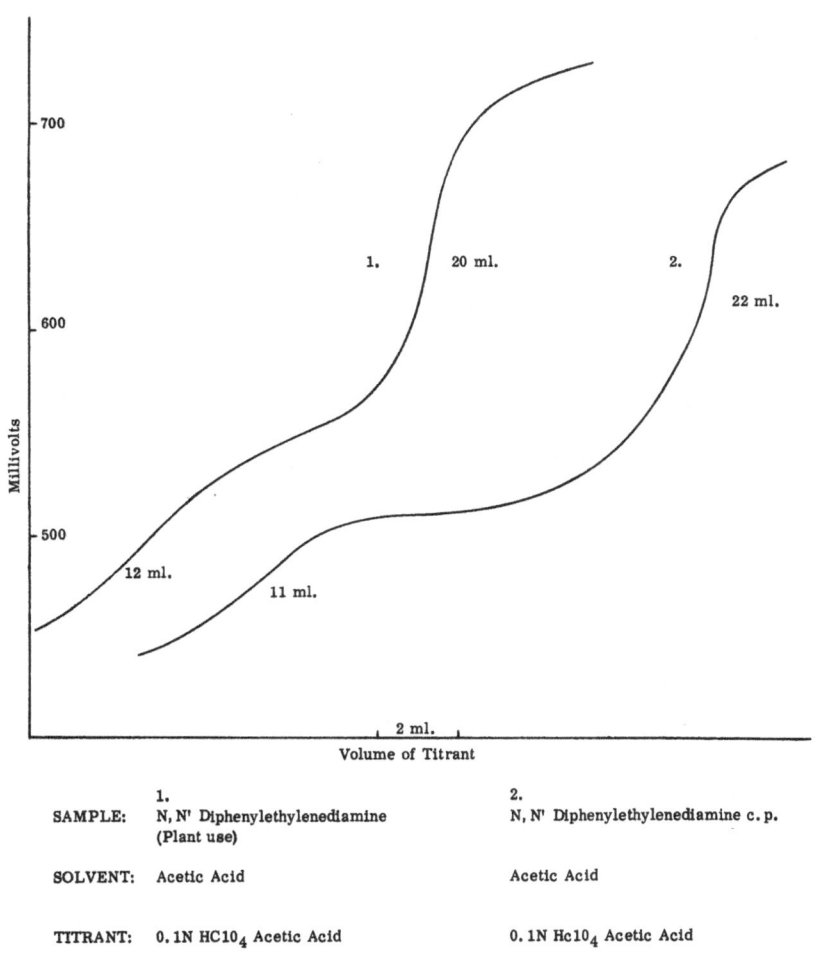

Fig. 3

Add a calculated amount of acetic anhydride to react with the water from the 72% acid. The solution is standardized against potassium acid phthalate.

Glacial acetic acid.

Acetic anhydride, reagent grade.

PROCEDURE

For a tertiary amine determination, weigh a sample containing up to 1 meq of 1,4-diphenylpiperazine into a 150-ml beaker. Add 20 ml of acetic anhydride and heat for 2 hr on a hot plate at 70°C. Cool the sample, add 20 ml of acetic acid, and titrate with 0.1 N perchloric acid.

For a characteristic titration of 1,4-diphenylpiperazine from which the N,N'-diphenylethylenediamine has been removed, weigh 0.2 g of sample in a 150-ml beaker. Add 40 ml of acetic anhydride and titrate with 0.1N perchloric acid. One equivalent of amine per mole is titrated in the tertiary amine determination and two equivalents of amine per mole are titrated in the characteristic titration for 1,4-diphenylpiperazine.

RESULTS

Recovery of 1,4-diphenylpiperazine from a N,N'-diphenylethylenediamine sample, previously found to contain 13.75% of the compound, was 100.1% in the tertiary amine determination. Recovery from a precipitated sample was 98.07%.

CONCLUSIONS

Unknown tertiary amines would be included in the direct titration. Separation of the N,N'-diphenylethylenediamine and water-soluble amines will help in showing the presence of other impurities.

REFERENCES

1. Fiegl, F. "Spot Tests in Organic Analysis," (Elsevier Publishing Co.) 5th ed., p. 215.
2. Ciaccio, L. L., Missan, S. R., McMullen, W. H., and Grenfall, T. C. Anal. Chem., Vol. 29, p. 1671, 1957.